Advances in Anatomy
Embryology and Cell Biology

Vol. 91

Editors

F. Beck, Leicester W. Hild, Galveston
R. Ortmann, Köln J.E. Pauly, Little Rock
T.H. Schiebler, Würzburg

Agnes A.M. Gribnau
Leonardus G.M. Geijsberts

Morphogenesis of the Brain in Staged Rhesus Monkey Embryos

With 19 Figures

Springer-Verlag
Berlin Heidelberg New York Tokyo
1985

Dr. Agnes A.M. Gribnau
Leonardus G.M. Geijsberts

Department of Anatomy and Embryology,
Faculty of Medicine, University of Nijmegen,
P.O. Box 9101, 6500 HB Nijmegen, The Netherlands

ISBN-13:978-3-540-13709-2   e-ISBN-13:978-3-642-69953-5
DOI: 10.1007/978-3-642-69953-5

Library of Congress Cataloging in Publication Data
Gribnau, Agnes Antoinette Marie.
Morphogenesis of the brain in staged Rhesus monkey embryos.
(Advances in anatomy, embryology, and cell biology; v. 91)
Includes bibliographical references and index.
1. Brain. 2. Developmental neurology. 3. Morphogenesis. 4. Rhesus monkey--Develop-
ment. 5. Embryology--Mammals.
I. Geijsberts, L.G.M., 1927–   . II. Title. III. Series.
QL801.E67 vol. 91 [QL801] 574.4 s [599.8'2] 84-14020
ISBN-13:978-3-540-13709-2 (U.S.)

© Springer-Verlag Berlin Heidelberg 1985

The use of general descriptive names, trade names, trade marks, etc. in this publication, even if the former are not especially identified, is not to be taken as a sign that
such names, as understood by the Trade Marks and Merchandise Marks Act, may
accordingly be used freely by anyone.
Product Liability: The publisher can give no guarantee for information about drug
dosage and application thereof contained in this book. In every individual case the
respective user must check its accuracy by consulting other pharmaceutical literature.

2121/3140-543210

Contents

Contents

Abbreviations

The following abbreviations are used in the figures

Ad	adenohypophysis		
Ait	adhesio interthalamica	Is	isthmus
Am	amygdaloid complex	I_1, I_2	isthmic neuromeres
ap	alar plate	LP	lens placode
BO	bulbus olfactorius	LR	lateral recess
bp	basal plate	LV	lens vesicle
CA	commissura anterior	lvr	lateral ventricular ridge
CeF	cervical flexure	Mam	mamillary region
CER	cerebellum	MES	mesencephalon
CGL	corpus geniculatum laterale	mvr	medial ventricular ridge
ch	choroid plexus	MYEL	myelencephalon
CI	capsula interna	M_1, M_2	mesencephalic neuro-
cip	crista interparencephalica		meres
CO	chiasma opticum	Neu	neurohypophysis
CP	commissura posterior	nI, nII,	nerve I, nerve II, etc.
cp	cerebellar plate	etc.	
CR	corpus restiforme	Oi	oliva inferior
CrF	cranial flexure	OP	olfactory placode
CS	corpus subthalamicum	OPi	olfactory pit
	Luysi	Os	oliva superior
DI	diencephalon	OS	optic stalk
DT	dorsal thalamus	OV	otic vesicle
ep	epiphysis	PAR a	parencephalon anterius
ET	epithalamus	PAR p	parencephalon posterius
F	fornix	PC	primary cortex
FL	frontal lobe	PF	pontine flexure
FM	foramen Monroi	PoO	postoptic region
FMT	fasciculus mamillotegmen-	PrO	preoptic region
	talis	PROS	prosencephalon
FR	fasciculus retroflexus	PrRuT	prerubral tegmentum
GnV,	ganglion nervi V, VII, and	PrT	pretectal region
VII,VIII	VIII	Pu	putamen
GP	globus pallidus	RD	retinal disc
hcc	hypothalamic cell cord	rec l	recessus lateralis
HYP	hypothalamus	Rh_1-Rh_7	rhombomeres 1-7
Hyp a	anterior hypothalamic re-	RHOMB	rhombencephalon
	gion	SC	spinal cord
Hyp p	posterior hypothalamic re-	sdb	sulcus diencephalicus
	gion		basalis

sdd	sulcus diencephalicus dorsalis	ssi	sulcus subpallii intermedius
sdm	sulcus diencephalicus medius	ST	subthalamus
		st	sulcus terminalis
sdv	sulcus diencephalicus ventralis	std	sulcus telodiencephalicus
		Str	striatum
sia	sulcus intraencephalicus anterior	SYN	synencephalon
		TEL	telencephalon
sih	sulcus interhemisphericus	Th	torus hemisphericus
sli	sulcus lateralis infundibuli	TL	temporal lobe
s lim	sulcus limitans of His	TO	tractus opticus
SM	stria medullaris	Tr	trigeminal nucleus
smb	sulcus mesencephalicus basalis	Tt	torus transversus
		TS	tractus solitarius
sml	sulcus mesencephalicus lateralis	Ve	vestibular nuclei
		ve	ventricular eminence
SO	supraoptic region	VT	ventral thalamus
srb	sulcus rhombencephalicus basalis	zl	zona limitans intrathalamica
srl	sulcus rhombencephalicus lateralis	I	lateral ventricle
		III	third ventricle
ssd	sulcus subpallii dorsalis	IV	fourth ventricle

1 Introduction

During development two strongly interrelated processes can be discerned in the central nervous system (CNS), namely morphogenesis and histogenesis. Most neuroembryological studies deal with histogenetic features virtually without any morphological elucidation. It must be stressed, however, that histogenetic investigations should be based upon a thorough knowledge of morphogenesis. This holds especially for the forebrain, which during development is subjected to drastic transformations, particularly when only two-dimensional sections are used. Therefore the present study on morphogenesis forms the first part of a research project on the ontogenesis of the brain in the rhesus monkey. The second part (Gribnau and Geijsberts 1984) will deal with the early histogenesis of the forebrain.

The first recognizable precursor of the CNS in vertebrates is the neural plate, which, after the formation of the germ layers, is induced in the ectoderm. The lateral margins of the neural plate start to rise, forming a neural groove. Eventually, they meet dorsally in the midline and fuse, resulting in the formation of the neural tube. The ultimate sites of closure at either end of the neural tube are called the anterior and posterior neuropores. Before the closure of the anterior neuropore, which precedes that of the posterior neuropore, the anlage of the CNS can be divided into a narrow elongated caudal part, the future spinal cord, and a wider rostral part, the precursor of the brain. Within the latter part, von Kupffer (1906) discerned two vesicle-like structures which he termed the archencephalon (rostral) and the deuterencephalon (caudal). Soon after the appearance of these two vesicles the deuterencephalon becomes divided into two parts: the mesencephalon and the rhombencephalon. From that moment on the archencephalon is called the prosencephalon. The prosencephalon and the rhombencephalon each give rise to two vesicles. Thus at that stage the brain consists of five consecutive compartments, named, from rostral to caudal: telencephalon, diencephalon, mesencephalon, metencephalon, and myelencephalon (Huxley 1871).

The subdivision of the prosencephalon occurs at the time that the telencephalic hemispheres arise as lateral evaginations of the prosencephalon. The most rostral part of the wall of the prosencephalon does not participate in the evagination process and persists as the telencephalon medium up to the adult stage. The prosencephalic lumen thus becomes subdivided into the diencephalic part, which is the third ventricle; the two lateral ventricles of the cerebral hemispheres; and the lumen belonging to the telencephalic medium, which is called the ventriculus impar. The lateral ventricles maintain an open communication with the third ventricle via the two interventricular foramina of Monro.

Initially, the entire CNS consists of a thin wall enclosing a wide ventricle. During further development a thickening of the wall occurs, coupled with a narrowing of the ventricular lumen, especially in the basal parts of the brain.

In the different individual areas of the brain, however, this process occurs neither at the same time nor at the same rate: the development of the brain proceeds in a heterochronous manner. This heterochrony accounts to some extent for the morphogenetic transformations of the brain during its development. The underlying heterochronous histogenetic processes will be dealt with in a following paper (Gribnau and Geijsberts 1984).

Another morphogenetically important aspect of the developing CNS is the progressive impact of the cerebral flexures on the morphology of the brain. Even before the closure of the neural tube the anlage of the brain is characterized by a dorsally convex curvature at the border between the future prosencephalon and mesencephalon. This flexure was called flexura cephalica by His (1888), plica encephali ventralis by von Kupffer (1906), and flexura cranialis by Bartelmez and Evans (1926); the latter term will be used in the present investigation. A second dorsally convex flexure originates at the junction between rhombencephalon and spinal cord: the flexura cervicalis. Later, a third, but ventrally convex, curvature arises at the rhombencephalic level: the flexura pontina. Thus the morphogenesis of the CNS is also marked by the development of these three curvatures in the neuraxis.

In the classic concept of the vertebrate CNS, which is based upon the ontogenetic work of His (1893) and the comparative studies of Herrick (1899) and Johnston (1902), among others, as reviewed by Nieuwenhuys (1974), the lower brain stem consists of four functional longitudinal zones separated by ventricular sulci. The basal plate consisting of two motor columns is separated from the alar plate, consisting of two sensory columns, by the sulcus limitans of His. On the basis of a comparative study on the amphibian brain, Herrick (1910) also concluded the presence of four longitudinal zones in the diencephalon. He subdivided this part of the brain into hypothalamus, pars ventralis thalami, pars dorsalis thalami, and epithalamus, in basal to dorsal order. Like Herrick, Kuhlenbeck (1929a, b) considered the four longitudinal zones delimited by ventricular sulci to be the basic elements of the structural plan of the CNS in vertebrates. Later, Kuhlenbeck (1930, 1933, 1936, 1954), on the basis of comparative and embryological studies, also provided strong evidence for the existence of this longitudinal zonal pattern in the CNS of birds and mammals, including man.

In contrast with this concept of longitudinal zones is the subdivision of the embryonic brain into transversely oriented subunits, as was advocated by Bergquist (1932), Källén (1952, 1953, 1955), Källén and Lindskog (1953), Bergquist and Källén (1954), and Vaage (1969). These subunits or neuromeres, being a number of consecutive bulges oriented transversely to the neuraxis, can be recognized during early development as described earlier by Orr (1887), von Kupffer (1906), and Meek (1907). The Swedish authors were even able to discern three successive waves of neuromeres: the proneuromeres in the open neural tube, the (secondary) neuromeres in the closed neural tube, and finally, the tertiary or postneuromeres. The latter structures are characterized by the presence of so-called migration areas, or subdivisions of which the constituent cells have a high mitotic activity in contrast to the bordering zones which have low mitotic activities. The postneuromeres gradually fade away in favor of the migration areas, which can be traced throughout development and ultimately give rise to the various brain nuclei.

The question of whether the transversely oriented neuromeres or the longitudinal zones should be considered the fundamental morphological subunits of the brain was posed by Keyser (1972). In his comprehensive study on the development of the diencephalon of the Chinese hamster, this author provided substantial evidence for the neuromere theory. The two concepts, however, might be less incompatible than they seem.

In the telencephalon the neuromeric pattern is absent, as was also concluded by Bergquist and Källén (1955), and the morphogenetic process is dominated by quite another phenomenon. After their origin as lateral evaginations of the thin-walled prosencephalon, the telencephalic hemispheres expand dorsally, frontally, and caudally. Their basolateral parts soon start to thicken, forming bilateral intraventricular protrusions. During further development the ventricular elevations on both sides are replaced by longitudinal ventricular ridges — a medial one and a lateral one. These two ventricular ridges gradually merge into one ventricular eminence which ultimately gives rise to the structures constituting the striatum. The process as briefly indicated here was extensively described in the Chinese hamster by Lammers et al. (1980), who also made it reasonable to assume that the lateral ventricular ridge has a purely telencephalic origin, whereas the medial ventricular ridge is partly of diencephalic and partly of telencephalic nature. Secondly, these authors provided substantial evidence for the theory that the two ridges arise as separate structures, the medial one before the lateral one, as was also proposed by Kodama (1926), Grünthal (1952), Hewitt (1958, 1961), Brown (1967), and Kahle (1969). This opinion was contradictory to the view supported by some other authors (Hochstetter 1919; Källén 1951; Hamilton et al. 1972; among others) who suggested that the original intraventricular protrusion in both hemispheres becomes divided into a medial and a lateral part. In the present investigation the question of whether the findings of Lammers et al. (1980) in a rodent also apply to primates will be considered.

2 Materials and Methods

The material employed in the present investigation consists of the brains of rhesus monkey (*Macaca mulatta*) embryos, varying in postconceptional age from 28 ± 1 to 50 ± 1 days (E_{28}–E_{50}). The same material was used in a previous investigation (Gribnau and Geijsberts 1981), so for detailed information on the animals used, breeding method, pregnancy diagnosis, hysterotomy technique, estimation of embryonic age, collection and processing of the embryos, and the staging method, the reader is referred to that publication. Only the most essential information will be mentioned briefly here.

The animals used were kept in a colony consisting at any one time of 20 adult females and 3 adult males placed in individual cages and fed on a commercially prepared pellet diet with additional fruit and water ad libitum. A total of 64 females and 8 males were used. The menstrual cycle was recorded daily and in most specimens lasted 28 days, but varied between 27 and 31 days. At the time selected for mating (around ovulation time) the female was individually caged with the male during a 72-h period, starting on the morning of day 10 after the onset of menstrual bleeding.

Pregnancy was diagnosed using an immunological test based upon the detection of chorionic gonadotropin in the urine of pregnant animals, as described previously (Gribnau 1975), on postmating days 19 through 21. The average gestation period in our population was approximately 168 days.

The embryos were collected by hysterotomy under deep O_2, N_2O, halothane anesthesia. After removal of the embryos from the uterus, they were immediately fixed by immersion in Bouin's fluid (Romeis 1968) for at least 48 h. Thereafter the specimens were processed in alcohol (progressively ranging from 50% to 100%), methylbenzoate, amylacetate, and paraffin wax (melting point 56° C) according to a time schedule adapted to the size of the material. After embedding, all embryos were serially sectioned on a Spencer rotary microtome at 7 μm in one of the three conventional directions (horizontal, transverse, or sagittal). The sections were stained either with hematoxylin and eosin or Mayer's modification of the Nissl technique (Romeis 1968), or impregnated according to Bodian (1936), for cells and fibers respectively.

The lateral, frontal, dorsal, and superior aspects of the brains of all specimens were photographically recorded during the methylbenzoate phase, in which the embryos are translucent. Small embryos were photographed under a Zeiss operation microscope with a Zeiss/Ikon camera unit, whereas gross specimens were photographed with an Ihagee EXA 500 camera. Microscopic sections were photographed with either an automatic Zeiss Photomicroscop II or a Leitz Aristophot. Generally, Ilford Pan F film was used, except for the Aristophot pictures, for which Agfa Pan 25 Professional film appeared to be more suitable.

The embryonic age was calculated by counting the number of days (x) elapsed after the second day of mating. The latter day was considered to be postconceptional day 1 or embryonic day 1 (E_1). The error in the embryonic age resulting from the mating procedure used amounts to plus or minus 1 day, thus the age of the embryos will be indicated as $E_{x\pm1}$. In our analysis encompassing the whole organogenetic period embryos ranging in age from $E_{28\pm1}$ to $E_{50\pm1}$ were used.

As was demonstrated in our previous publication (Gribnau and Geijsberts 1981) neither embryonic age nor embryonic length is a suitable criterion to use in accurately defining successive developmental stages, since both appear to vary considerably. Thus a staging system for rhesus monkey embryos was introduced in which each stage was defined on the basis of external and internal morphological characteristics. The embryonic period was divided into 23 successive developmental stages according to the Carnegie system, previously used in human (O'Rahilly 1979) and baboon (Hendrickx 1971) embryos, but also proven to be applicable to rodents (Gribnau and Geijsberts 1981). Together, these 23 developmental stages constitute three successive phases during embryonic development: the presomite phase (stages 1–8), the somite phase (stages 9–12), and the postsomite or organogenetic phase (stages 13–23). A detailed description of the material used in the present investigation was given in our previous publication and is summarized in Table 1.

Three-dimensional reconstructions of the brains of stage 15, 17, 19, 20, and 23 embryos were made using the reconstruction technique developed by Gribnau and Lammers (1976), which is based upon a comparison of two different series of sections of two brains per developmental stage. These two series are sectioned both at right angles to the median plane as well as perpendicular to each other: horizontally and transversely. The reconstruction technique comprises, firstly, the graphical reconstruction of each of the two median sections of the brains of both specimens, using the orthogonal projection technique reviewed by Gaunt (1971), and secondly, the optimalization of both reconstructions by mutual comparison. Each resultant optimalized graphical reconstruction is then used as a reference in the correct piling of the three-dimensional reconstruction of the respective series.

Table 1. Material used in the present study

Number of embryos	Developmental stage	Embryonic age (in days p.c.)[a]	Embryonic length (mm)
5	13	(27) 28–30 (33)	4.5–6
5	14	(27) 30–32 (33)	6–8
2	15	(29) 30–33 (35)	7–9
3	16	(31) 32–34 (37)	7–11
6	17	(32) 34–36 (38)	9–12
5	18	(34) 35–38 (39)	11–15
8	19	(35) 36–42 (43)	14–17
5	20	(35) 38–42 (45)	16–20
4	21	(39) 40–44 (45)	18–22
3	22	(43) 44–48 (49)	20–25
6	23	(45) 46–50 (51)	24–30

[a] Minimum and maximum ages found are shown in parentheses

Every tenth section of the series was traced on a 3.8-mm-thick polystyrene sheet, at such a magnification that the thickness of one sheet corresponded with the interval between two successively traced sections (10×7) multiplied by the magnification. The tracings were cut out using a hot wire and then superimposed, using the optimalized graphical reconstruction of the median section of the brain in question as a reference. Drawings were made after these polystyrene models, showing (a) the lateral aspect of the brain, (b) the medial aspect of the brain after the model was cut in the median plane, and (c) the superior aspect of the forebrain after removal of parts of the diencephalic and hemispheric walls.

3 Description of the Stages

During the presomite stages in mammals, the fertilized egg (stage 1) develops into the embryonic disc with a primitive streak and an incipient neural plate (stage 8). During the somite stages the neural plate stage with 0–3 pairs of somites (stage 9) develops into the neural tube stage with 21–29 pairs of somites (stage 12). During the postsomite or organogenetic stages, the embryo with more than 30 pairs of somites and a closing or closed neural tube (stage 13) develops into the embryo in which all organs and organ systems are present and the secondary palate is closing (stage 23). In mammals the transition from embryo into fetus, marking the end of stage 23, is defined as the time of closure of the secondary palate, as reviewed by Gribnau and Geijsberts (1981).

In all species analyzed so far, both primates and rodents, the anterior neuropore invariably closes during stage 11. The closure of the posterior neuropore, however, occurs during stage 12 in primates and during stage 13 in rodents. Apart from this exception great conformity exists in the literature on the descriptions of the early development of the CNS during the somite stages. For that reason the present investigation on the morphogenesis of the CNS of the rhesus monkey starts at the beginning of the organogenetic phase, stage 13.

The morphogenetic process of the brain was studied not only with microscopic sections but also with the aid of three-dimensional reconstructions of brains from embryos at a number of developmental stages, namely the organogenetic stages 15, 17, 19, 20, and 23, as described in the preceding section.

3.1 Stage 13

In stage 13 embryos the CNS of the rhesus monkey consists of a thin-walled neural tube in which both the anterior and posterior neuropore are closed. Within the neural tube the prosencephalon, mesencephalon, and rhombencephalon can be identified in the embryos cleared in methylbenzoate (Fig. 1 A). The cervical and cranial flexures are easily recognizable, both causing clear bends in the neuraxis. Although the prosencephalon shows a dorsal evagination which is the primordial telencephalic hemisphere, a subdivision into telencephalon and diencephalon cannot be made since a telodiencephalic boundary can hardly be indicated.

In the microscopic sections of the CNS of a stage 13 embryo the thin-walled character of the neural tube is clearly illustrated (Fig. 1 B–E). In the prosencephalon the primordial telencephalic and diencephalic areas can be indicated as shown by Fig. 1 B and D, respectively. The telodiencephalic boundary, however, cannot be identified, since a broad transition area is present which can

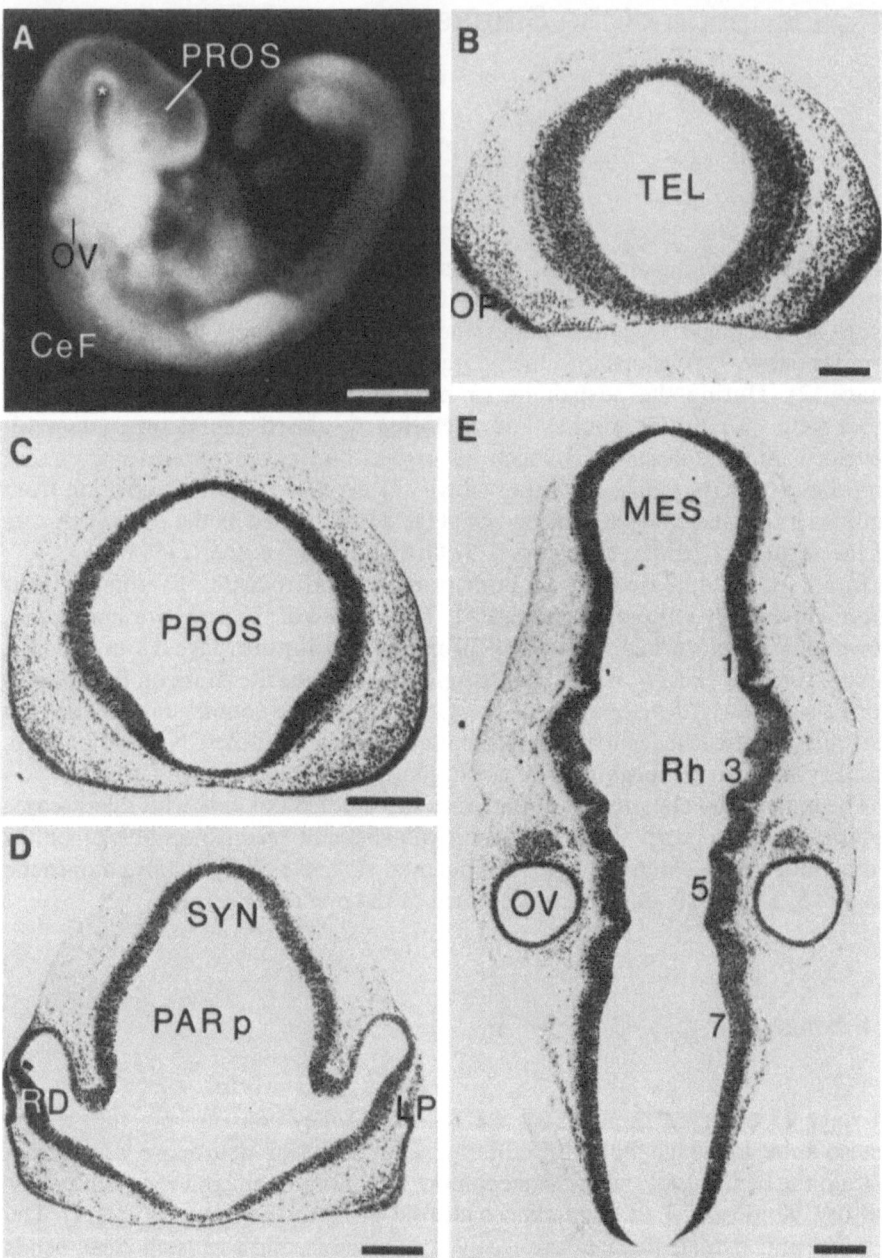

Fig. 1. Photomicrographs of stage 13 embryos, aged 32 ± 1 days p.c. *A* Lateral view of the embryo cleared in methylbenzoate; the *asterisk* is situated in the cranial flexure; *calibration bar* = 1.0 mm. *B, C, D,* and *E* Frontal sections; *calibration bars* = 0.2 mm

not be assigned to either the telencephalon or the diencephalon (Fig. 1 C). In the future diencephalic area of the prosencephalon the two most caudal neuromeres, namely the synencephalon and the parencephalon posterius, can be recognized (Fig. 1 D). The olfactory placodes, lens placodes, and retinal discs characteristic of stage 13 embryos are illustrated (Fig. 1 B, D). Caudal to the synencephalon the mesencephalon is situated, in which only a vague indication of a subdivision into two neuromeres could be recognized in horizontal sections. Caudally, the mesencephalon borders upon the isthmus region. In the rhombencephalic part of the CNS two phenomena dominate the morphology: firstly the very thin roof of the fourth ventricle and secondly, the presence of pronounced metameric structures in its basal part, which are the rhombomeres (Fig. 1 A, E). A total of seven consecutive rhombomeres could be identified; the otic vesicle, which is closed in stage 13 embryos, being opposite to the fifth one (Fig. 1 E). The primordial ganglia of the cranial nerves V, VII, VIII, X, and XII could be identified.

3.2 Stage 14

The outer form of the CNS in stage 14 embryos has slightly changed as compared with that in the preceding stage (e.g., Figs. 1 A, 2 A). Between the deep cranial flexure and the cervical flexure, which forms approximately a 90° angle, an incipient pontine flexure has developed. Secondly, a further dorsal expansion of the telencephalic area is clearly illustrated. From the microscopic sections it can be deduced that the evagination of the telencephalic hemispheres has not only proceeded dorsally but also laterally (Fig. 2 B). A sulcus interhemisphericus is still absent in stage 14 embryos. A deep sulcus telodiencephalicus marks the boundary between the dorsal and lateral parts of the telencephalon and the diencephalon (Fig. 2 A, B). Frontally, however, this sulcus fades away, thus the subdivision of the prosencephalon in its frontal part is still incomplete.

In the frontal part of the diencephalon a subdivision into three longitudinal zones is present, caused by the sulcus diencephalicus basalis and the sulcus diencephalicus ventralis (Fig. 2 B). These two sulci delimit a slightly thickened area, which is the future hypothalamic cell cord. The deeply indented optic cup, characteristic of stage 14 embryos, communicates freely with the third ventricle through a wide optic stalk (Fig. 2 C). The morphology of the caudal part of the diencephalon is characterized by the presence of its transversely oriented subunits or neuromeres, the parencephalon posterius and the synencephalon, which can even be seen in the methylbenzoate pictures (Fig. 2 A). The parencephalon posterius is obviously wider than the synencephalon; the frontal and caudal boundaries of both neuromeres consist of annular constrictions.

Caudal to the synencephalon, the mesencephalon is situated, in which a vague annular constriction marks the boundary between the two mesencephalic neuromeres, m_1 and m_2 (Fig. 2 A). The mes-rhombencephalic boundary is easier to recognize as the constriction of the future isthmus area (Fig. 2 A, D). The two characteristics of the rhombencephalon described for stage 13, namely the very thin roof of the fourth ventricle and the presence of rhombomeres in

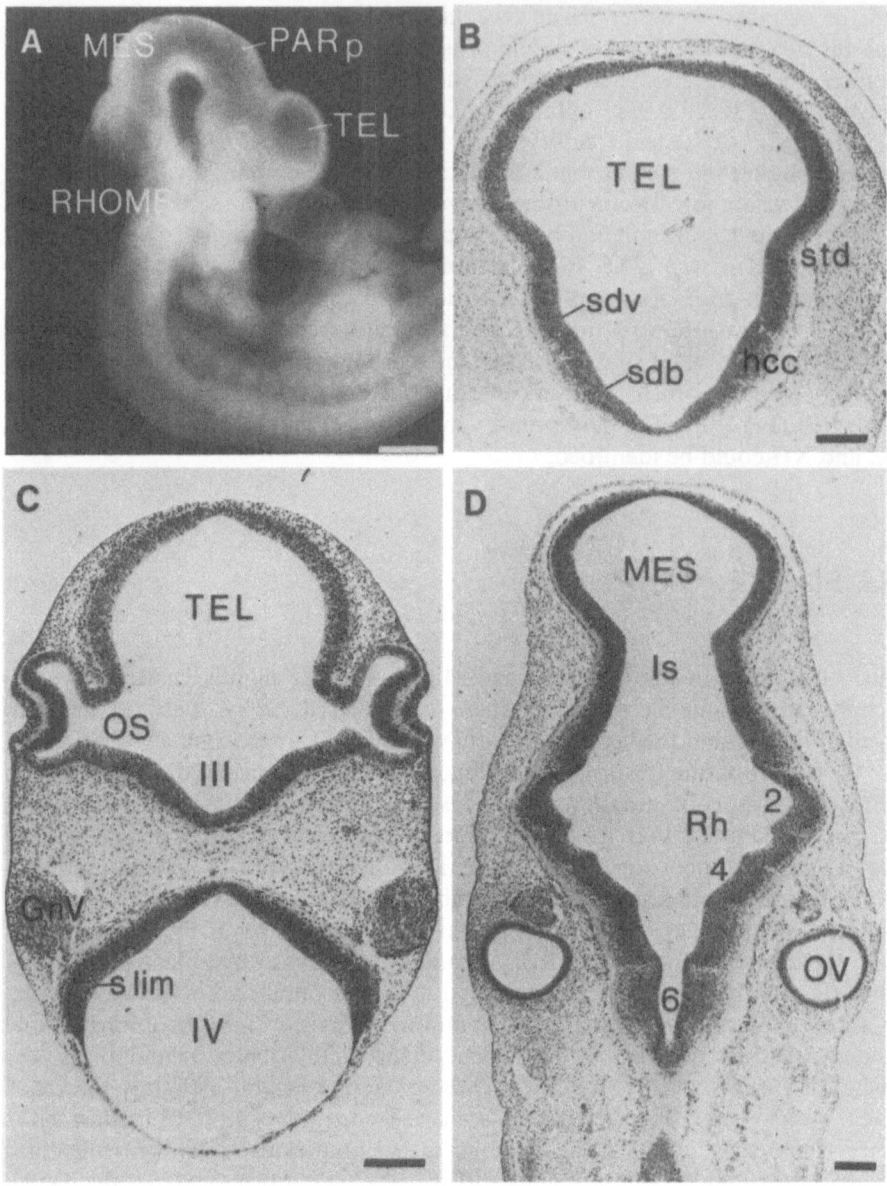

Fig. 2. Photomicrographs of stage 14 embryos, aged 30 ± 1 days p.c. *A* Lateral view of the embryo cleared in methylbenzoate; *calibration bar* = 1.0 mm. *B* and *C* Horizontal sections, *D* frontal section; *calibration bars* = 0.3 mm

its basal part, are still present (Fig. 2A, C, D). The seven rhombomeres are very distinctive; the second is the widest, and the fifth is located opposite to the otic vesicle (Fig. 2D). In addition a shallow sulcus limitans is present (Fig. 2C). The ganglia of the cranial nerves are now massive structures (Fig. 2C, D).

3.3 Stage 15

In stage 15 embryos the evagination of the anlage of the telencephalic hemispheres has proceeded, particularly in the dorsal and lateral directions (Fig. 3A). The flexura pontina has slightly deepened as compared with the preceding stage (e.g., Figs. 2A, 3A). Although all parts of the CNS have grown, the outer form of the developing brain has hardly changed. In broad outline the same applies to the inner form, since the neural tube is still thin-walled in virtually all its parts, as can be deduced from the microscopic sections (Fig. 3B, C, D), thus causing the outer surface to be nearly parallel to the inner surface. A groove on the outside, therefore, corresponds with a ridge on the inside, as can also be deduced from the pictures of the three-dimensional reconstructions (Fig. 4A, B, C). The latter phenomenon is clearly illustrated by the sulcus telodiencephalicus on the outside and its counterpart, the torus hemisphericus ridge, on the inside. Both structures, which can also be seen in the sections (Fig. 3B, C), now completely separate the telencephalon from the diencephalon. Thus, as far as the forebrain is concerned, developmental stage 15 is a very crucial one in which the division of the prosencephalon into its subunits, the telencephalon and the diencephalon, is completed.

The telencephalon consists of two primordial hemispheres, which caudally and laterally form two distinct bulges but dorsally and frontally merge into a single vesicle, which, close to the median plane, is extremely thin-walled (Figs. 3A, B, C, 4A, B, C). In the frontal part of the telencephalon a sulcus interhemisphericus is still absent. The sulcus telodiencephalicus on the outside and the torus hemisphericus on the inside, starting from the lamina terminalis, curve around the anlage of the hemisphere in front of the optic stalk towards the torus transversus frontally (Figs. 3A, 4A, B). On the inside the wide, presumptive interventricular foramen of Monro can be seen, bordered basally by the torus hemisphericus. The telencephalon consists completely of a thin-walled neuroepithelium (Fig. 3B, C).

In the frontal part of the diencephalon the hypothalamic cell cord is more prominent now than in the preceding stage because of a further thickening of its wall (Figs. 3B, 4B). As a consequence the two grooves forming the boundary, the sulcus diencephalicus basalis and the sulcus diencephalicus ventralis, are accentuated. The latter sulcus runs frontally into the broad optic stalk, which opens into the optic cup (Figs. 3C, 4B). Characteristic of stage 15 embryos are both the closure of the lens vesicle (Fig. 3C) as well as the presence of the first retinal pigment (Fig. 3A). Frontally, the medial boundary of the diencephalon consists of the following structures in a dorsal to basal order: the torus transversus, the recessus preopticus, the connection between the left

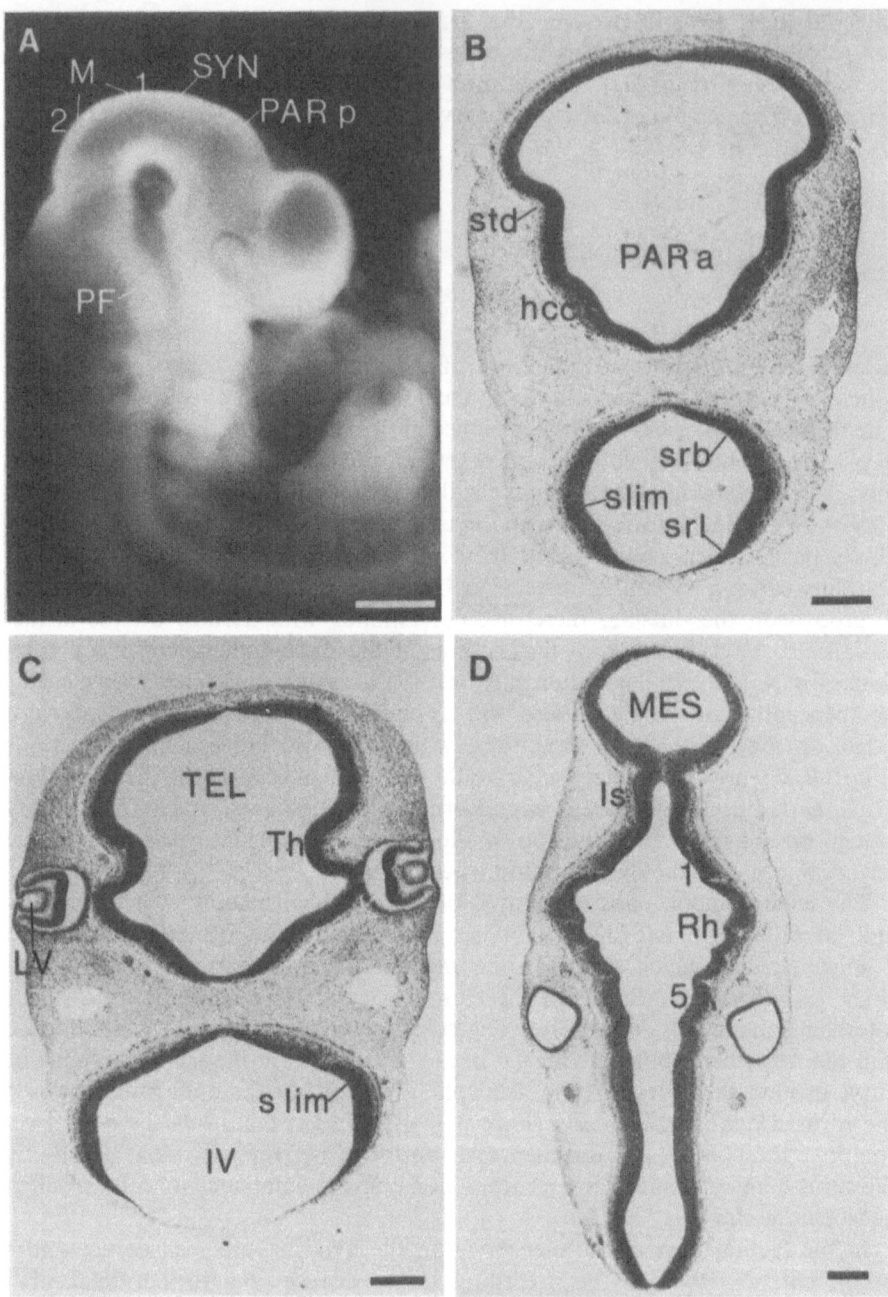

Fig. 3. Photomicrographs of stage 15 embryos, aged 30 ± 1 days p.c. *A* Lateral view of the embryo cleared in methylbenzoate; *calibration bar* = 1.0 mm. *B* and *C* Horizontal sections, *D* frontal section; *calibration bars* = 0.3 mm

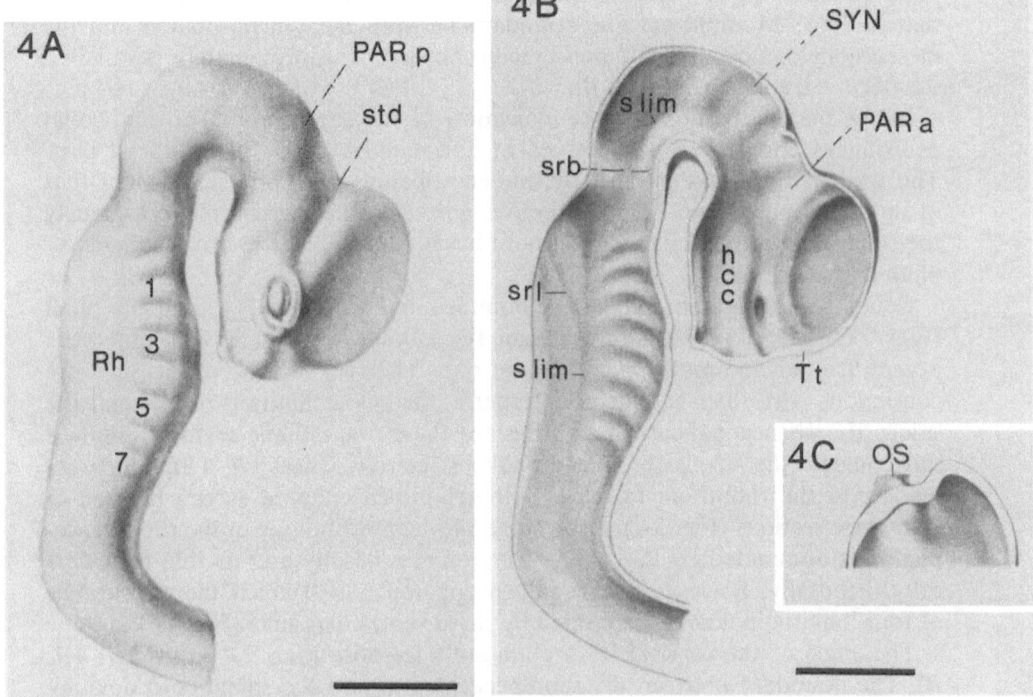

Fig. 4. Three-dimensional reconstructions of the brain of a stage 15 embryo, aged 30±1 days p.c. *A* Lateral view, *B* medial view, *C* superior view of the prosencephalon after removal of parts of the telencephalon and the diencephalon; *calibration bars* = 1.0 mm

and right hypothalamic cell cord, and, finally, the recessus postopticus (Fig. 4*B*). In the floor of the diencephalon the primordial infundibular recess can be recognized frontally.

Between the torus hemisphericus and the sulcus diencephalicus ventralis a thin-walled area is situated, which is caudally bounded by the rostral border of the parencephalon posterius (Figs. 3*B*, 4*B*). This area, forming an excavation on the inside (Fig. 4*B*) and a bulge on the outside (Fig. 4*A*), represents the dorsal part of the parencephalon anterius. From later stages it could be deduced that this area will generate the ventral thalamus during further development. The annular constriction between the two parencephalic neuromeres on the outside (Fig. 4*A*) and its counterpart, the crista interparencephalica, on the inside (Fig. 4*B*) can easily be identified. The parencephalon posterius is a little wider than the parencephalon anterius in front of it, and the synencephalic neuromere caudal to it. The parencephalon posterius will give rise during further development to the dorsal thalamus, among other things. The boundaries of the most caudal diencephalic neuromere, the synencephalon, are also very distinct (Fig. 4*A, B*). The basal parts of the parencephalon posterius and the synencephalon have a slightly wider wall than their dorsal parts (Fig. 4*B*). It can be stated that the morphology of the diencephalon is characterized by horizontal zones in its frontal part and by transversely oriented neuromeres in its caudal

13

part. This pattern of the diencephalon in stage 15 embryos strongly resembles that in stage 14 embryos. The boundary between the synencephalon and the mesencephalon, or the di-mesencephalic border, is approximately parallel to the flexura cranialis (Fig. 4A, B).

In the mesencephalon a faint indication of the presence of two individual neuromeres could only be observed in horizontal sections of its dorsal part. The wall of the basal part of the mesencephalon is slightly wider than that of its dorsal part. Two sulci can be recognized in the mesencephalon, namely the sulcus limitans of His and the more basally located sulcus basalis mesencephali (Figs. 3D, 4B).

Caudally, the mesencephalon is bounded by the isthmus rhombencephali (Figs. 3D, 4A, B). In the isthmus region, two isthmic neuromeres were observed: a wide frontal one and a small caudal one. The two mesencephalic sulci are continuous with their isthmic counterparts, the sulcus limitans of His and the sulcus rhombencephalicus basalis. Besides these, the isthmic region exhibits a third sulcus: the sulcus rhombencephalicus lateralis (Figs. 3B, 4B), which extends into the rhombomeric part of the rhombencephalon as can be seen in transverse sections (Fig. 3C). As in stage 14, the morphology of the rhombencephalon is dominated by the seven rhombomeres basally and its thin roof dorsally. Frontally, however, the rhombencephalon has attained the appearance of four longitudinal zones separated by three ventricular sulci (Fig. 4B).

The angle of the cervical flexure amounts to more than 90° (Figs. 3A, 4A, B). The boundary between the rhombencephalon and the spinal cord dorsally is marked by the transition of the extremely thin roof of the fourth ventricle into the thin-walled roof of the spinal cord.

3.4 Stage 16

In embryos representative of developmental stage 16 the outer form of the CNS is dominated by the deep flexura cranialis followed by the flexura pontina, which has slightly deepened as compared with the preceding stage, and the flexura cervicalis, which now approximates a right angle (Fig. 5A).

For the first time during its ontogenesis the telencephalon consists of two complete individual hemispheres, since the sulcus interhemisphericus now also separates the most frontal parts (Fig. 5A, B). The evagination of the hemispheres has markedly proceeded as compared with stage 15 brains, especially in the dorsal and lateral directions (e.g., Figs. 3, 5). The interventricular foramen of Monro is still very wide in its dorsal and caudal parts (Fig. 5C, D). In its basal and frontal parts, however, it is narrowed by the deepened sulcus telodiencephalicus on the one hand and by a local thickening of the telodiencephalic border zone on the other (Fig. 5C, D). This local thickening of the basal telodiencephalic wall is the first sign of the future medial ventricular ridge, as it was termed by Lammers et al. (1980), which in the literature is also called the medial striatal ridge.

In the diencephalon the neuromeres present are easily recognizable: the parencephalon anterius with the optic evagination, the parencephalon posterius, and the synencephalon (Fig. 5A, C). The two parencephalic neuromeres are sepa-

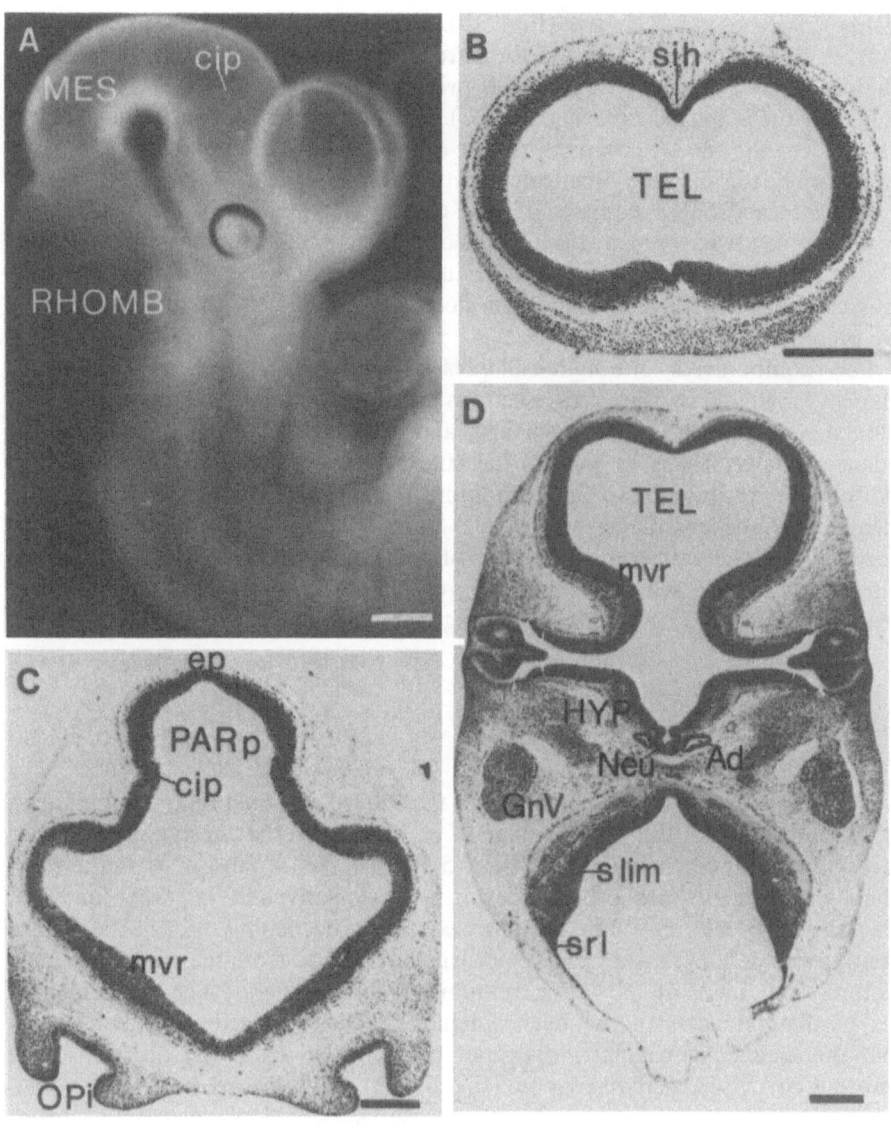

Fig. 5. Photomicrographs of stage 16 embryos, aged 32 ± 1 days p.c. *A* Lateral view of the embryo cleared in methylbenzoate; *calibration bar* = 0.5 mm. *B* and *C* Frontal sections, *D* horizontal section; *calibration bars* = 0.3 mm

rated by the crista interparencephalica. The optic stalk has elongated, but still shows a rather wide lumen which is continuous with the third ventricle (Fig. 5*D*). In the diencephalic floor the neurohypophysial evagination can easily be recognized, which together with the two adenohypophysial lobes of Rathke's pouch forms the hypophysial anlage (Fig. 5D). In the roof of the diencephalon, just in front of the junction between the parencephalon posterius and the synencephalon, a small second evagination is present medially, which is the primordial anlage of the epiphysis cerebri (Fig. 5C). The parencephalon posterius is markedly wider than the synencephalon. The syn-mesencephalic boundary can easily be recognized in the methylbenzoate picture (Fig. 5A).

The mesencephalon consists of two thin-walled mesencephalic neuromeres both having a wide lumen, the frontal one being smaller than the caudal one. The latter neuromere caudally borders upon the rhombencephalon at the isthmus rhombencephali, in which some decussating fibers of the trochlear nerve are present dorsally.

The rhombomeres, although still present in this stage, are less conspicuous than in stage 15. In transverse sections of the rhombencephalon (Fig. 5D) it can be seen that both the basal and alar plates have widened, whereas the sulcus limitans of His as well as the sulcus lateralis are both deepened. The alar plates constitute the so-called rhombic lips, in which during further development the cerebellar anlagen will arise. Caudally and laterally the rhombic lips border upon the extremely thin roof of the fourth ventricle (Fig. 5A, D). At the cervical flexure, which amounts approximately to a 90° angle, the rhombencephalon passes into the spinal cord (Fig. 5A).

3.5 Stage 17

Apart from a proportional growth of all the subunits of the brain, a comparison of the outer form of the CNS in stage 17 embryos with that in stage 16 embryos reveals only minor changes (e.g., Figs. 5A, 6A). The expansion of the telencephalic hemispheres has proceeded not only dorsally and laterally, but also frontally and occipitally (Fig. 6B). Moreover, the deepening of both the flexura pontina and the flexura cervicalis can easily be deduced from the methylbenzoate pictures: the angle of the latter amounts to less than 90°. The thickening of the basolateral parts of the telencephalic wall has proceeded in such a way that the medial ventricular ridges can be identified from the outside in the embryos cleared in methylbenzoate (Fig. 6A, B).

Fig. 6. Photomicrographs of stage 17 embryos, aged 35 ± 1 to 37 ± 1 days p.c. A and B Lateral and superior views of the embryo cleared in methylbenzoate; *calibration bars* = 1.0 mm. C Frontal section, D–G sagittal sections; *calibration bars* = 0.5 mm

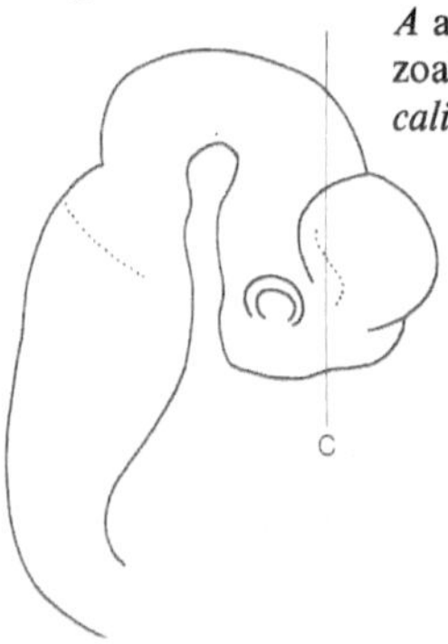

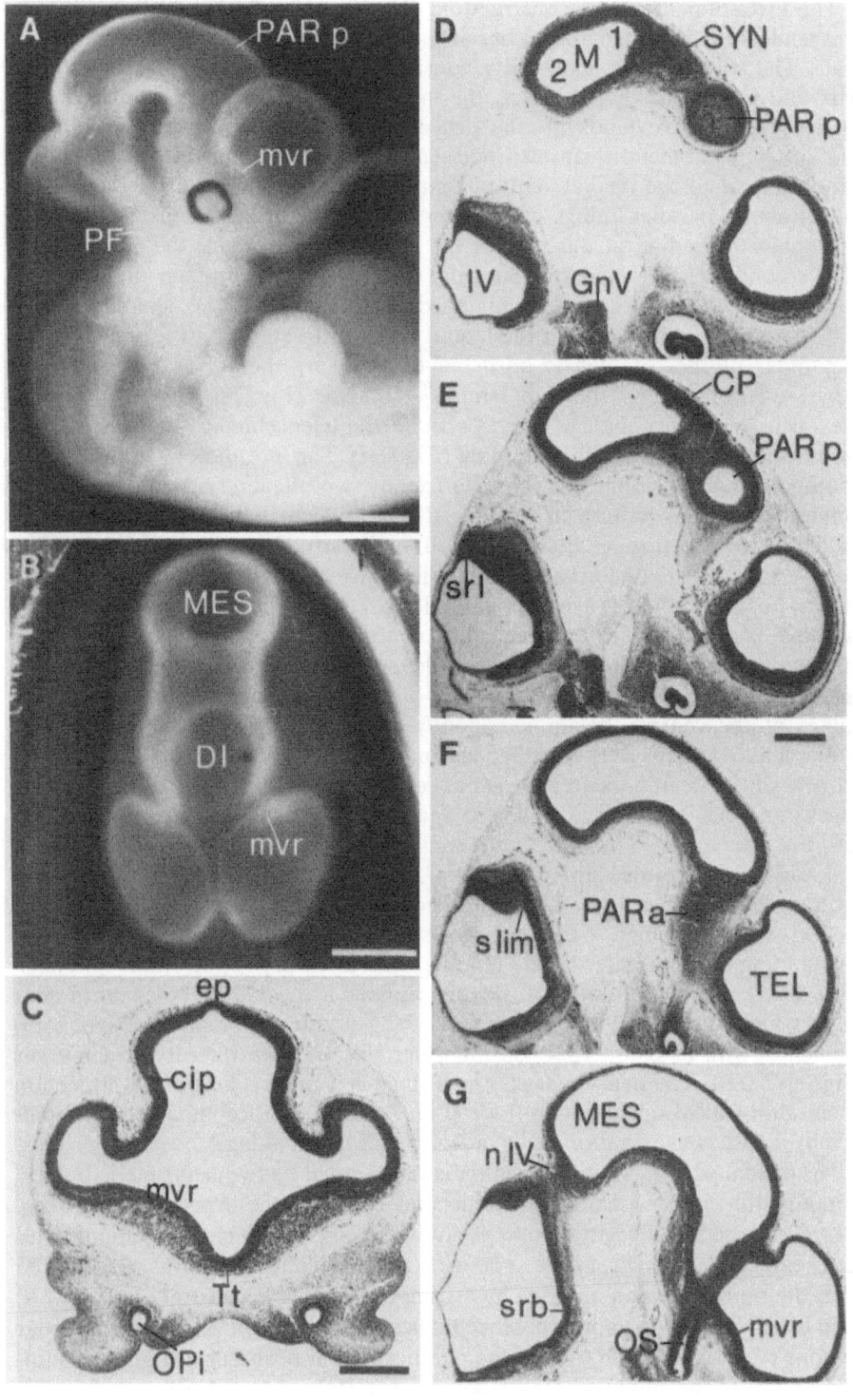

The three-dimensional reconstructions (Fig. 7) not only substantiate the general tendencies described above, but also elucidate transformations on a smaller scale. The telencephalic hemisphere consists of a dorsal thin-walled ballooning part and a wide basolateral part, the medial ventricular ridge, which via the foramen of Monro extends into the diencephalic hypothalamic wall (Fig. 7 B, C). The latter phenomenon can also be seen in the section illustrated in Fig. 6 C. Frontolateral to the medial ventricular ridge, part of the hemispheric wall no longer shows parallel linings of its inner and outer surfaces (Figs. 6 F, G, 7 C). The slight thickening of this part of the hemisphere indicates the development of a second or lateral ventricular ridge. Although the foramen of Monro is basally bounded by the expanding medial ventricular ridge, it still forms a wide communication between the lateral and third ventricles (Fig. 7 B, C). The remainder of the foramen is bordered caudally by the sulcus hemisphericus ridge and dorsally by the thin lamina terminalis (Fig. 7 B). On the outside, the sulcus telodiencephalicus curves around the telencephalic hemisphere, running frontally, dorsal to the optic stalk (Fig. 7 A). The morphology of the diencephalon of stage 17 embryos is characterized by two outstanding features: (1) the longitudinal zonal pattern in its frontal part, which is formed by two thick-walled rostrocaudal zones alternating with two rostrocaudal grooves; and (2) the transversely oriented neuromeric bulges in its caudal part, namely the parencephalon posterius and the synencephalon (Fig. 7 A, B). The two thick-walled longitudinal zones of the frontal diencephalon are represented basally by the hypothalamic cell cord; and dorsally by the diencephalic part of the medial ventricular ridge, frontally, and the dorsal part of the parencephalon anterius, or the anlage of the ventral thalamus, caudally. The latter two parts are separated by the shallow sulcus intraencephalicus anterior (Fig. 7 B). The two rostrocaudal grooves of the frontal diencephalon are constituted by the sulcus diencephalicus basalis and the sulcus diencephalicus ventralis, the latter fades away frontally into the optic stalk. The optic stalk itself exhibits a lumen over its full length and ends in the optic cup (Figs. 6 D–G, 7 B). The optic cup is characterized by a wide optic fissure and the presence of the lens, which in microscopic sections shows a lumen with a crescent shape. The latter feature was the most outstanding internal criterion in defining stage 17 embryos (Gribnau and Geijsberts 1981). The frontal part of the diencephalic area of the medial ventricular ridge, which during further development will generate the preoptic region, communicates with its counterpart on the other side through the torus transversus (Figs. 6 C, 7 B). The neurohypophysial anlage is a finger-like frontal projection of the thin-walled diencephalic floor (Fig. 7 A, B). It is accompanied on either side by a rostral evagination of the adenohypophysial anlage.

The caudal part of the diencephalon is represented by two neuromeric bulges: frontally the parencephalon posterius and caudally the synencephalon (Fig. 7 A, B). The series of sagittal sections shown in Fig. 6 D–G clearly demonstrates the presence of the neuromeres. The wide parencephalon posterius is separated from the parencephalon anterius by a transverse annular constriction (Fig. 7 A, B) in which a small crista interparencephalica can be recognized in perpendicular sections (Fig. 6 C). In the roof of the parencephalon posterius a slightly evaginated anlage of the epiphysis cerebri can be seen (Figs. 6 C, 7 A, B).

Caudally, the parencephalon posterius is succeeded by the synencephalon, the former neuromere being considerably wider than the latter (Figs. 6 D–G,

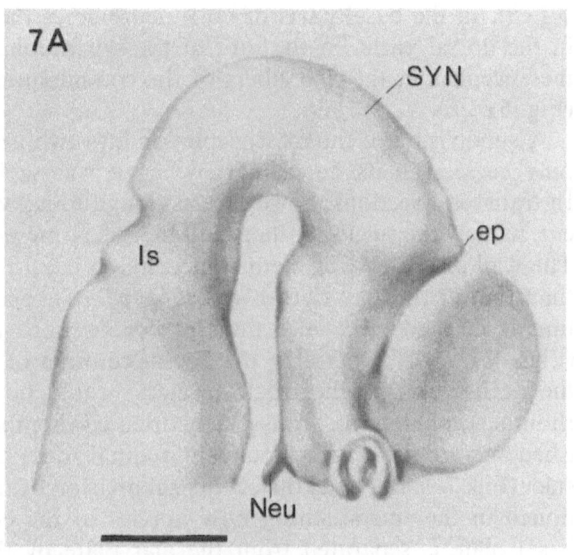

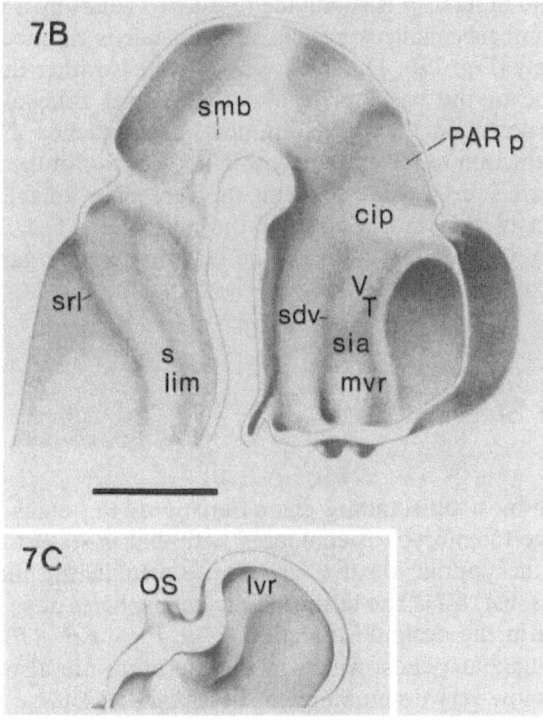

Fig. 7. Three-dimensional reconstructions of the brain of a stage 17 embryo, aged 34±1 days p.c. *A* Lateral view, *B* medial view, *C* superior view of the prosencephalon after removal of parts of the telencephalon and the diencephalon; *calibration bars* = 1.0 mm

$7A, B$). In the basal parts of both neuromeres the wall is slightly thicker than in the dorsal parts. In the roof of the synencephalon at the junction with the mesencephalon, the first fibers of the commissura posterior can be recognized (Fig. $6D, E$).

A subdivision of the mesencephalon into two neuromeres, m_1 and m_2, could only be seen in its dorsal part, when it was sagitally sectioned (Fig. $6D, E$). In transverse sections, however, two longitudinal ventricular sulci are still present, namely the sulcus limitans and the sulcus mesencephalicus basalis (Fig. $7B$). The wall of the part of the mesencephalon basal to the sulcus limitans is wider than that of the area dorsal to it (Fig. $6D–G$). The mes-rhombencephalic junction is marked by a deep annular constriction: the isthmus rhombencephali (Figs. $6A, D–G, 7A, B$). In the dorsal confines of the isthmus region fibers of the decussation of the trochlear nerve could be identified (Fig. $6G$). In the rhombencephalon, the transversely oriented rhombomeres have completely vanished and are replaced by four longitudinal zones separated by three ventricular sulci (Fig. $7A, B$). Thus the classic subdivision of the lower brain stem, as mentioned in the Introduction, now applies to the entire rhombencephalon: the basal plate is separated from the alar plate by the sulcus limitans and each plate in its turn is subdivided into two zones by a ventricular groove, the sulcus rhombencephalicus basalis and the sulcus rhombencephalicus lateralis, respectively (Fig. $7B$). The most ventral zone is rather thin-walled, whereas the dorsal zone of the basal plate and the ventral zone of the alar plate, which both border upon the sulcus limitans, have reached a substantial width in contradistinction to the extremely thin-walled roof of the fourth ventricle (Fig. $6D–G$). During further development the cerebellar anlage will originate in the frontal part of the ventral zone of the alar plate. The fourth ventricle still shows a very wide lumen, whereas the large trigeminal ganglion can be identified connected with the basal plate (Fig. $6D, E$).

3.6 Stage 18

The most outstanding characteristic of the outer morphology of the brain of stage 18 embryos as compared with that in stage 17 is the progressive deepening of the pontine flexure, as can be seen in the methylbenzoate pictures (e.g., Figs. $6A, 8A$). The telencephalic hemispheres have expanded both in the frontal and in the occipital direction (e.g., Figs. $6B, 8B$). Additionally, the metencephalic alar plates, which in the literature are also called the cerebellar plates, are now very dominating entities (Fig. $8A, C$).

The thickening of the basolateral part of the telencephalic hemisphere, which is evident in the methylbenzoate picture (Fig. $8A$), consists of two parts: the medial and lateral ventricular ridges (Fig. $8D$). The two ridges are separated by the sulcus subpallii intermedius, whereas the sulcus subpallii dorsalis forms the boundary between the lateral ventricular ridge and the thin-walled pallial part of the telencephalic hemisphere.

In the frontal part of the diencephalon the hypothalamic cell cord, which is bounded basally by the sulcus diencephalicus basalis and dorsally by the

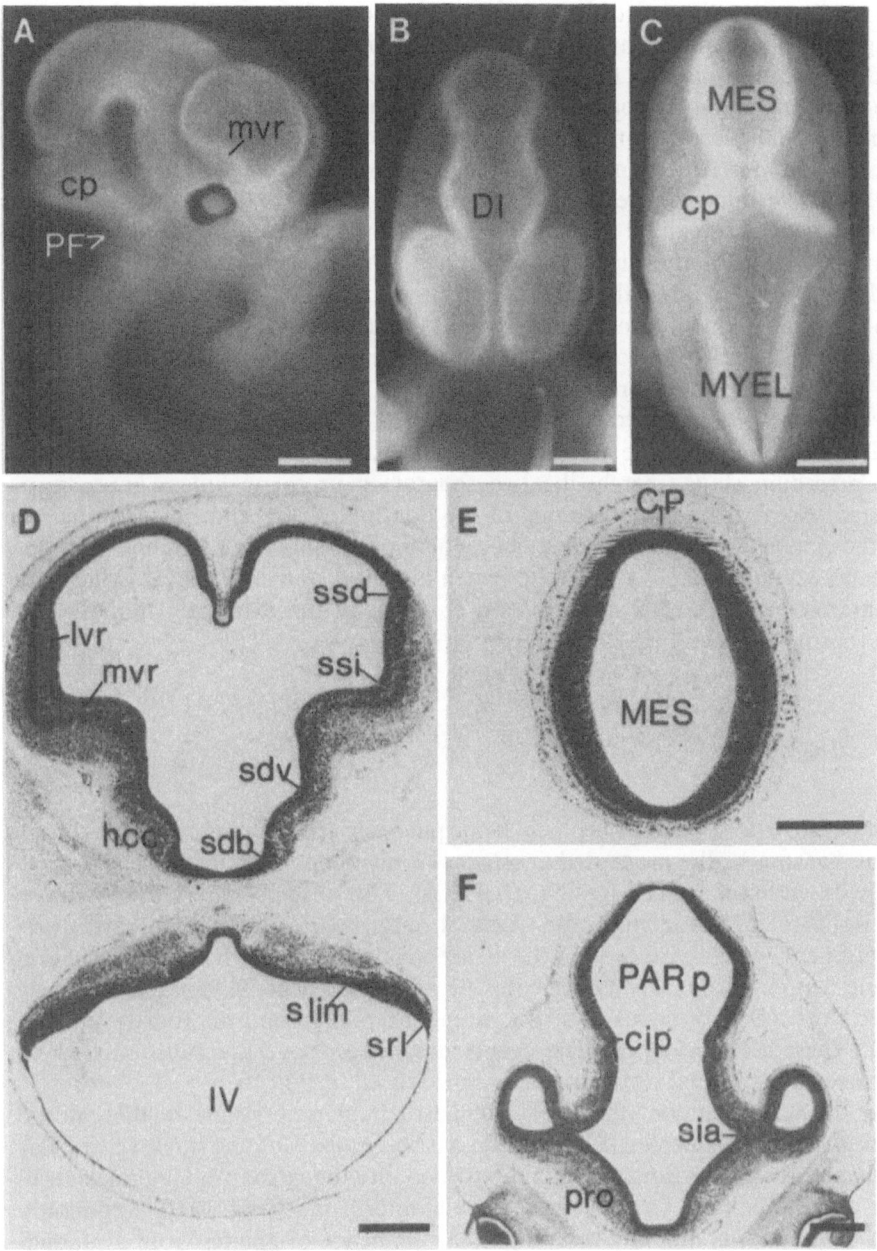

Fig. 8. Photomicrographs of stage 18 embryos, aged 35 ± 1 to 36 ± 1 days p.c. *A, B,* and *C* Lateral, superior, and dorsal views of the embryo cleared in methylbenzoate; *calibration bars* = 1 mm. *D* and *E* Horizontal sections, *F* frontal section; *calibration bars* = 0.5 mm

sulcus diencephalicus ventralis, protrudes far into the lumen of the third ventricle (Fig. 8D). The component parts of the dorsal diencephalon can easily be identified in a section parallel to the floor of the diencephalon (Fig. 8F). Frontally, the preoptic region is separated from the parencephalon anterius by the sulcus intraencephalicus anterior. Caudal to the parencephalon anterius the wide parencephalon posterius can be observed, the two neuromeres being separated by the crista interparencephalica. The most frontal part of the synencephalon can be recognized caudal to the parencephalon posterius. Looking at the developmental state of the diencephalic wall, a frontal-to-caudal gradient can be deduced from Fig. 8F: the wall of the preoptic region is wider than that of the parencephalon anterius, which in turn is wider than that of both the parencephalon posterius and the synencephalon. The number of fibers belonging to the commissura posterior, which is situated in the roof of the synencephalon, has considerably increased (Fig. 8E).

The morphology of the mesencephalon has hardly changed as compared with the preceding stage, merely the two separate neuromeric bulges have disappeared. Apart from the deepening of the pontine flexure and the accentuation of the cerebellar plates, the rhombencephalon also shows only minor morphogenetic changes (Fig. 8A, C). The widening basal and alar plates, which are separated by the sulcus limitans, are covered by the extremely thin roof of the fourth ventricle (Fig. 8D).

3.7 Stage 19

The outgrowth of the telencephalic hemispheres in the frontal and caudal directions has markedly proceeded in stage 19 in comparison with the brain of stage 18 embryos (e.g., Figs. 8A, B, 9A, B). The former process results in the development of the frontal pole. A basal outgrowth of the caudal part of the hemisphere, which partly covers the diencephalon, points to the initial formation of the temporal pole. The pontine flexure angle is now ca. 90°, which obviously affects the morphology of both the rhombencephalon and the fourth ventricle (e.g., Figs. 8A, C, 9A, C). The events described above are substantiated by the three-dimensional reconstructions reproduced in Fig. 10.

In the telencephalon, the thickened basolateral part visible in the embryo cleared in methylbenzoate (Fig. 9A, B) consists caudally of one single ventricular eminence. Rostrally, however, it is subdivided into the medial and lateral ventricular ridges separated by the sulcus subpallii intermedius (Fig. 10C). Apparently also in the rhesus monkey the ventricular eminence originates from a gradual process of merging of the two ventricular ridges, as was previously described by Lammers et al. (1980) in the Chinese hamster. This process, which starts caudally, gradually proceeds in the frontal direction. Because of the outgrowth of the ventricular eminence, the interventricular foramen of Monro has markedly narrowed (Fig. 10B), although in its frontal part it is still rather wide (Fig. 9D). The plexus choroideus of the lateral ventricle has started to develop. The anlage of the olfactory bulb is represented by a thickening of the frontal telencephalic wall, whereas the first olfactory fibers, coming from the nasal epithelium, have reached the brain (Figs. 9D, 10A, B).

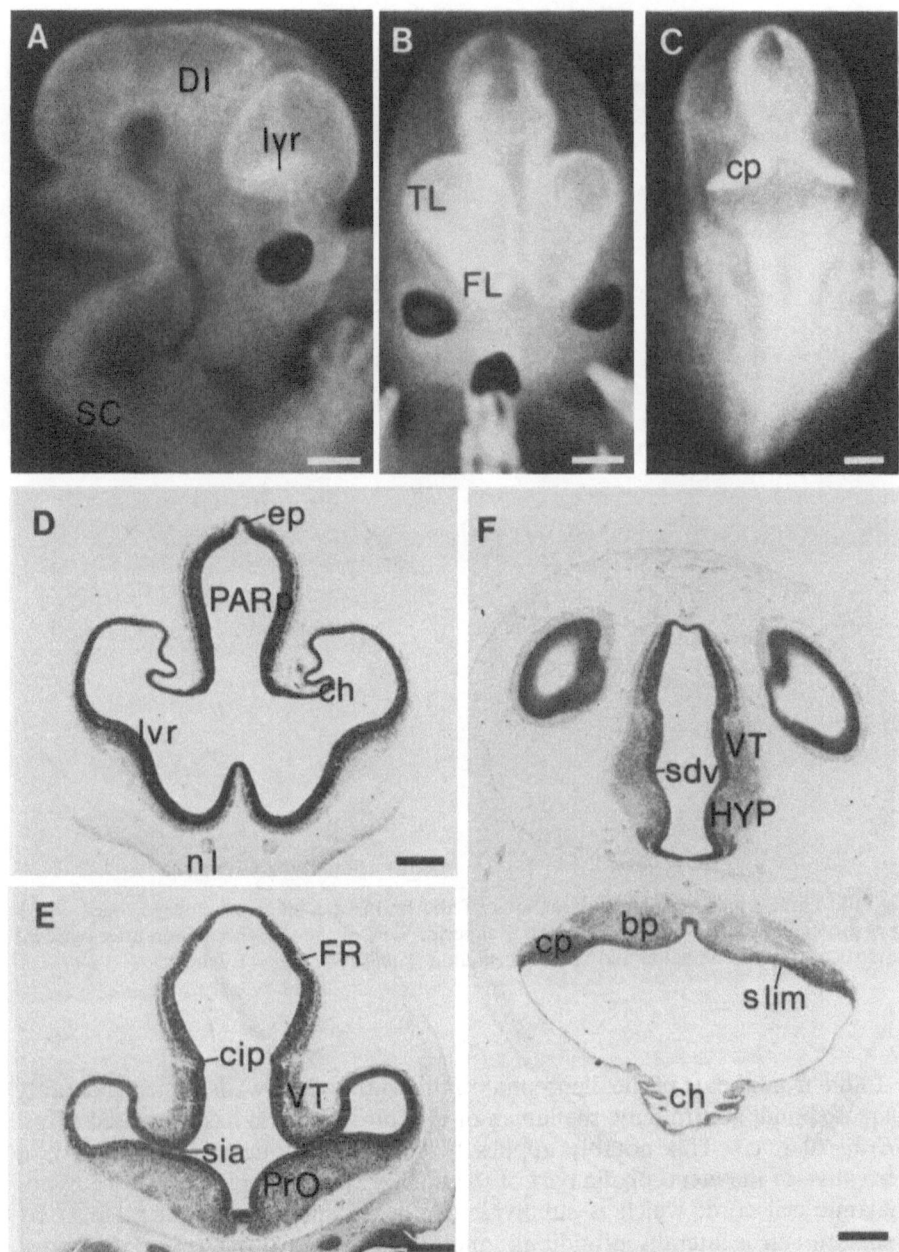

Fig. 9. Photomicrographs of stage 19 embryos, aged 36 ± 1 to 40 ± 1 days p.c. *A, B,* and *C* Lateral, ventral, and dorsal views of the embryo cleared in methylbenzoate; *calibration bars* = 1.0 mm. *D* and *E* Frontal sections, *F* horizontal section; *calibration bars* = 0.5 mm

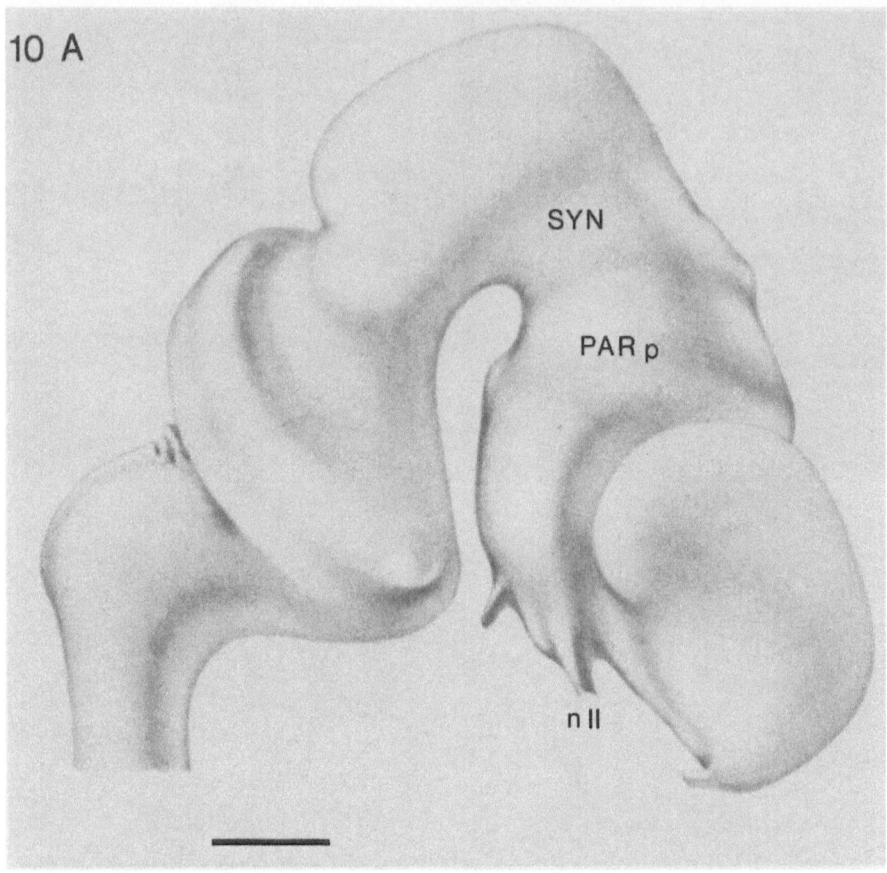

Fig. 10. Three-dimensional reconstructions of the brain of a stage 19 embryo, aged 36 ± 1 days p.c. *A* Lateral view, *B* medial view, *C* superior view of the prosencephalon after removal of parts of the telencephalon and the diencephalon; *calibration bars* = 1.0 mm

In the frontal part of the diencephalon the width of the walls has considerably increased and, accordingly, the lumen of the third ventricle has decreased (Figs. 9*E, F*, 10*B, C*). This notably applies to (1) the preoptic region, which is a derivative of the diencephalic part of the medial ventricular ridge; (2) the hypothalamic cell cord, which is subdivided into a frontal and a caudal part by a shallow sulcus lateralis infundibuli; and (3) the anlage of the ventral thalamus, which originates from the dorsal part of the parencephalon anterius (Fig. 10*B*). The preoptic regions of both sides communicate over the midline through the torus transversus (Fig. 9*E*). The sulcus intraencephalicus anterior, curving from the foramen of Monro towards the optic stalk, forms the caudobasal boundary of the preoptic region (Figs. 9*E*, 10*B*). The optic recess is established because of the closure of the optic stalk. Basal to the optic recess the frontal part of the thickened hypothalamic cell cord is located, which eventually will generate the supraoptic and postoptic regions (Fig. 10*B*). This area communicates with its counterpart on the opposite side by a broad junction in which during later

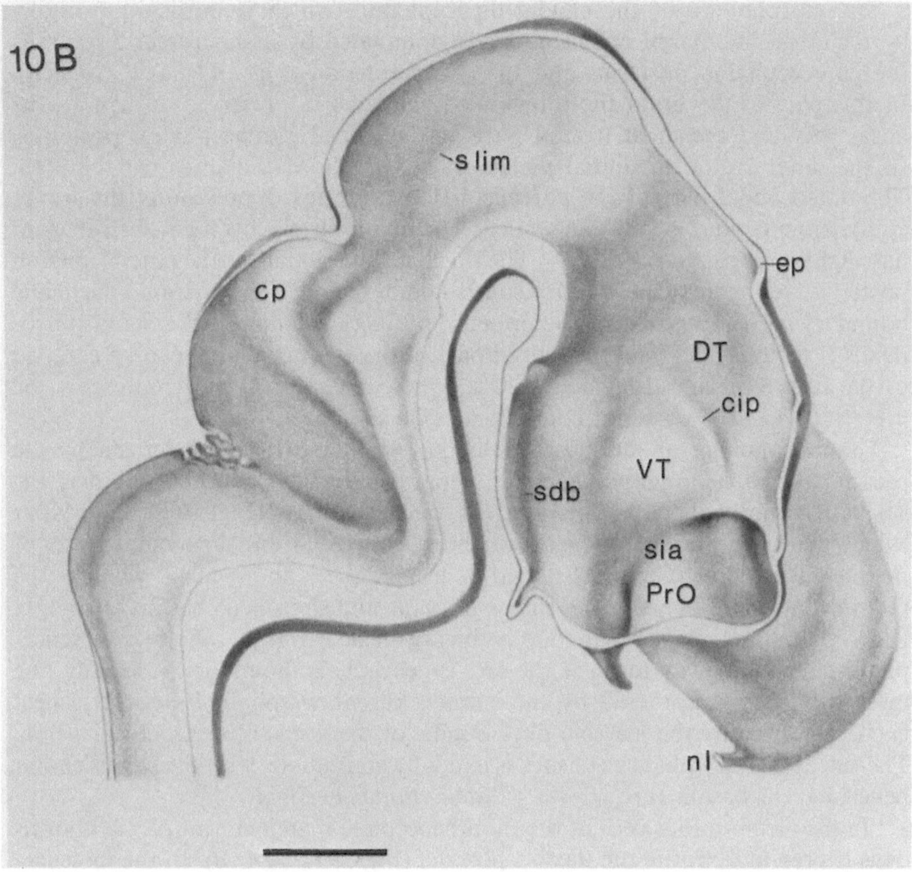

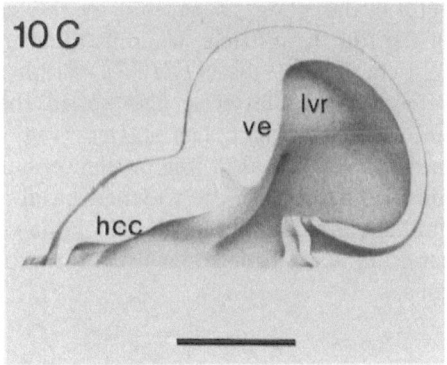

10 B

s lim

cp

ep

DT

cip

VT

sdb

sia
PrO

nl

10 C

lvr

ve

hcc

Fig. 10 B and C

stages the optic chiasm will develop. The neurohypophysis consists of a simple finger-like evagination of the floor of the frontal diencephalon, its lumen being continuous with the third ventricle. The caudal part of the thickened hypothalamic cell cord, located basal to the developing ventral thalamus, will give rise to the mamillary region among other things (Figs. 9 F, 10 B).

The morphology of the caudal diencephalon, which is bordered frontally by the crista interparencephalica, is still dominated by its neuromeric subunits the parencephalon posterius and the synencephalon (Figs. 9D, E, F, 10A, B). In this part of the brain the thin-walled character is preserved, except for its basal part and the most frontal part of the dorsal parencephalon posterius. In the latter region an initial widening of the dorsal thalamus can be seen. The dorsocaudal part of the parencephalon posterius, representing the future epithalamic region, is still thin-walled. The epiphysis cerebri located in its roof has slightly evaginated (Fig. 9D). The thickened basal part of the parencephalon posterius represents the future subthalamic region. The syn-parencephalic boundary is characterized by the presence of fibers forming the fasciculus retroflexus (Fig. 9E). The thickened basal part of the synencephalon can be equated to the future prerubral tegmental area, whereas its dorsal part represents the pretectal area. The posterior commissure is located in its roof.

The morphology of the mesencephalon is in stage 19 characterized by the presence of three longitudinal zones separated by two ventricular sulci: the sulcus mesencephalicus basalis and the sulcus limitans (Fig. 10B). The latter sulcus marks the boundary between the future tectal and tegmental parts of the mesencephalon. The most dorsal or tectal zone is thin-walled, whereas the intermediate zone between the sulcus limitans and the sulcus basalis is slightly wider. The third, most basal zone is thick-walled. Frontally, the two mesencephalic sulci fade away towards the mes-synencephalic boundary. Caudally, the mesencephalon is bordered by the isthmus rhombencephali; its dorsal aspect is characterized by the massive fiber bundle of the decussating trochlear nerve. The sulcus mesencephalicus basalis is caudally continuous with the sulcus rhombencephalicus basalis through the isthmus rhombencephali.

In the membranous roof of the rhombencephalon an incipient plexus choroideus is present opposite the flexura pontina (Figs. 9F, 10A, B). In the metencephalon, the four longitudinal zones are accentuated in comparison with previous stages. The ventral basal plate has slightly thickened; the dorsal basal and ventral alar plates, separated by the sulcus limitans, protrude into the lumen of the fourth ventricle, whereas the dorsal alar plate has retained its extremely thin character (Figs. 9F, 10B). Within the basal and alar plates − the dorsal subunit of the latter is also called the cerebellar plate − the formation of the nuclear anlagen has started (Fig. 9F). Although less pronouncedly than in the metencephalon, the myelencephalon is similarly characterized by its four longitudinal zones: the widened basal and alar plates, separated by the sulcus limitans, are dorsally bordered by the very thin roof (Fig. 10B). Via the flexura cervicalis these four zones are continuous with their counterparts in the spinal cord.

3.8 Stage 20

The progressive outgrowth of the telencephalic hemispheres by stage 20 has resulted in the formation of the frontal and temporal lobes (Figs. 11A, B, 12A, B). The temporal lobe laterally covers the dorsofrontal part of the diencephalon, whereas the frontal lobe reaches far beyond the most frontal level of the diencephalon. The primordial olfactory bulb is slightly detached from the frontal lobe, while the olfactory fibers now form a rather massive bundle (Fig. 12 A, B). The angles of both the pontine and the cervical flexure grow more and more acute, thereby the cerebellar plates gradually overhang the myelencephalon (Figs. 11 A, C, 12 A, B). Near the pontine flexure the cranial nerves V, VII, and VIII are present, emerging from the brain (Fig. 12 A).

In the methylbenzoate picture the ventricular eminence can be recognized from the outside (Fig. 11 A). A comparison of the two reconstructions of stages 19 and 20 reveals that the merging of the medial and lateral ventricular ridges into one ventricular eminence has proceeded frontally (e. g., Figs. 10 B, C, 12 B, C). Frontally, however, the two ventricular ridges are still separated by the sulcus subpallii intermedius, as is also shown in the section illustrated in Fig. 11 D. The outgrowth of the ventricular eminence has reduced the foramen of Monro to a narrow cleft (Figs. 11 E, 12 B, C). Basally, the ventricular eminence is continuous with the preoptic region, although the torus transversus and its lateral continuation mark the boundary between the two regions and thus form the telodiencephalic boundary. In the cerebral stem area the capsula interna can be identified (Fig. 11 F). Both the ventricular eminence caudally and the lateral ventricular ridge rostrally are dorsally bordered by the sulcus subpallii dorsalis, which separates them from the thin-walled primordial pallium (Figs. 11 D, E, 12 C). In the basolateral part of the pallial anlage a narrow superficial cortical layer (primary cortex) can be seen, consisting of darkly stained cells (Fig. 11 D). The plexus choroideus extends far into the lumen of the lateral ventricle.

The morphology of the diencephalon is no longer characterized by the presence of neuromeres: they have completely vanished (Fig. 12 A, B). The diencephalic wall now consists of more or less thick-walled parts separated by ventricular sulci (Fig. 12 B). The preoptic region constitutes the most frontal part of the diencephalon as was indicated earlier. This region is bounded dorsally by the lateral continuation of the torus transversus and caudobasally by the sulcus intraencephalicus anterior, which, coming from the foramen of Monro dorsally, curves around the preoptic region towards the optic recess (Figs. 11 F, 12 B). Fibers of the optic chiasm can readily be identified, crossing the median plane between the postoptic regions of both sides. Caudobasal to the optic chiasm the hypophysial evagination is present in the diencephalic floor. It shows hardly any developmental progress as compared with stage 19 (e. g., Figs. 10 B, 12 B). The sulcus lateralis infundibuli of earlier stages now has the appearance of a depression, which actually separates the anterior and posterior hypothalamic regions (Fig. 12 B). The latter region borders upon the mamillary region, in the floor of which the mamillary anlage can be recognized. The basal diencephalic regions just mentioned, namely the postoptic, the anterior and posterior hypothalamic, and the mamillary regions, are the derivatives of the basal part of the parencephalon anterius. Dorsally, these regions are bounded by the sulcus diencephalicus ventralis, which separates them from the primordial ventral thalamus, which in its turn is a derivative of the dorsal part of the parencephalon anterius (Fig. 12 B). The ventral thalamic anlage is rather thick-walled. In transverse sections (Fig. 11 F) the small ridge of the crista interparencephalica between the parencephalon anterius and the parencephalon posterius can easily

27

be recognized, and it thus represents the boundary between the developing ventral and dorsal thalamus. The ridge, coming from a point dorsally near the anlage of the plexus choroideus of the third ventricle, curves in a ventrally convex way towards the sulcus diencephalicus ventralis in the direction of the supramamillary recess. As a ridge it gradually fades away in its basal part, but histologically a clear natural boundary can be seen in the diencephalic wall between the parencephalon anterius and the parencephalon posterius (Fig. 11 F).

Frontally, the dorsal wall of the parencephalon posterius, in which the dorsal thalamus originates, has markedly increased in thickness, whereas its caudal part is still thin-walled (Figs. 11 F, 12 B). It may be stated that the development of the dorsal thalamus starts frontally and spreads caudally. Comparing the reconstruction of stage 20 with that of stage 19 (Figs. 10 B, 12 B), it appears that the ventricular part of the diencephalon is now mainly occupied by the dorsal thalamus at the expense of the ventral thalamus. The sulcus diencephalicus dorsalis forms the dorsocaudal boundary between the developing dorsal thalamus and the thin-walled epithalamic region, which also originates from the dorsal part of the parencephalon posterius. In the roof of the latter neuromere, the epiphysis cerebri is located, which has further evaginated as compared with the preceding stage (e. g., Figs. 10 A, B, 12 A, B). Basally, the sulcus diencephalicus ventralis separates the dorsal thalamus from the subthalamic region, which is the basal representative of the posterior parencephalic neuromere (Fig. 12 B). Morphologically, the boundary between the parencephalon posterius and the synencephalon can hardly be indicated, but the fasciculus retroflexus represents an outstanding landmark (Fig. 11 E, F). The synencephalon is thin-walled, except for its most basal part, which during further development gives rise to the prerubral tegmentum. Its dorsal part represents the regio pretectalis of later stages. The commissura posterior, which marks the synmesencephalic boundary, has markedly increased in volume (Fig. 11 F, H). The sulcus mesodiencephalicus marks the dimesencephalic boundary on the outside (Fig. 12 A).

In the mesencephalon the wall-thickening process has considerably advanced as compared with the preceding stage; in addition, a third sulcus, the sulcus mesencephalicus lateralis, has developed (e. g., Figs. 10 B, 11 G, H, 12 B). The ventral part of the basal plate, which is dorsally bounded by the sulcus mesencephalicus basalis, has markedly widened, but the most advanced area in this respect is the dorsal part of the basal plate (Figs. 11 G, H, 12 B). The ventral

Fig. 11. Photomicrographs of stage 20 embryos, aged 40±1 days p.c. *A, B*, and *C* Lateral, superior, and dorsal views of the embryo cleared in methylbenzoate; *calibration bars* = 1.0 mm. *D, E, F*, and *G* Frontal sections; *H* horizontal section; *calibration bars* = 0.5 mm

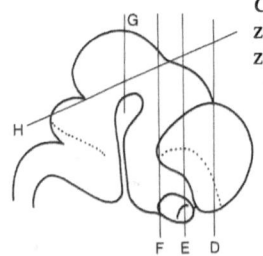

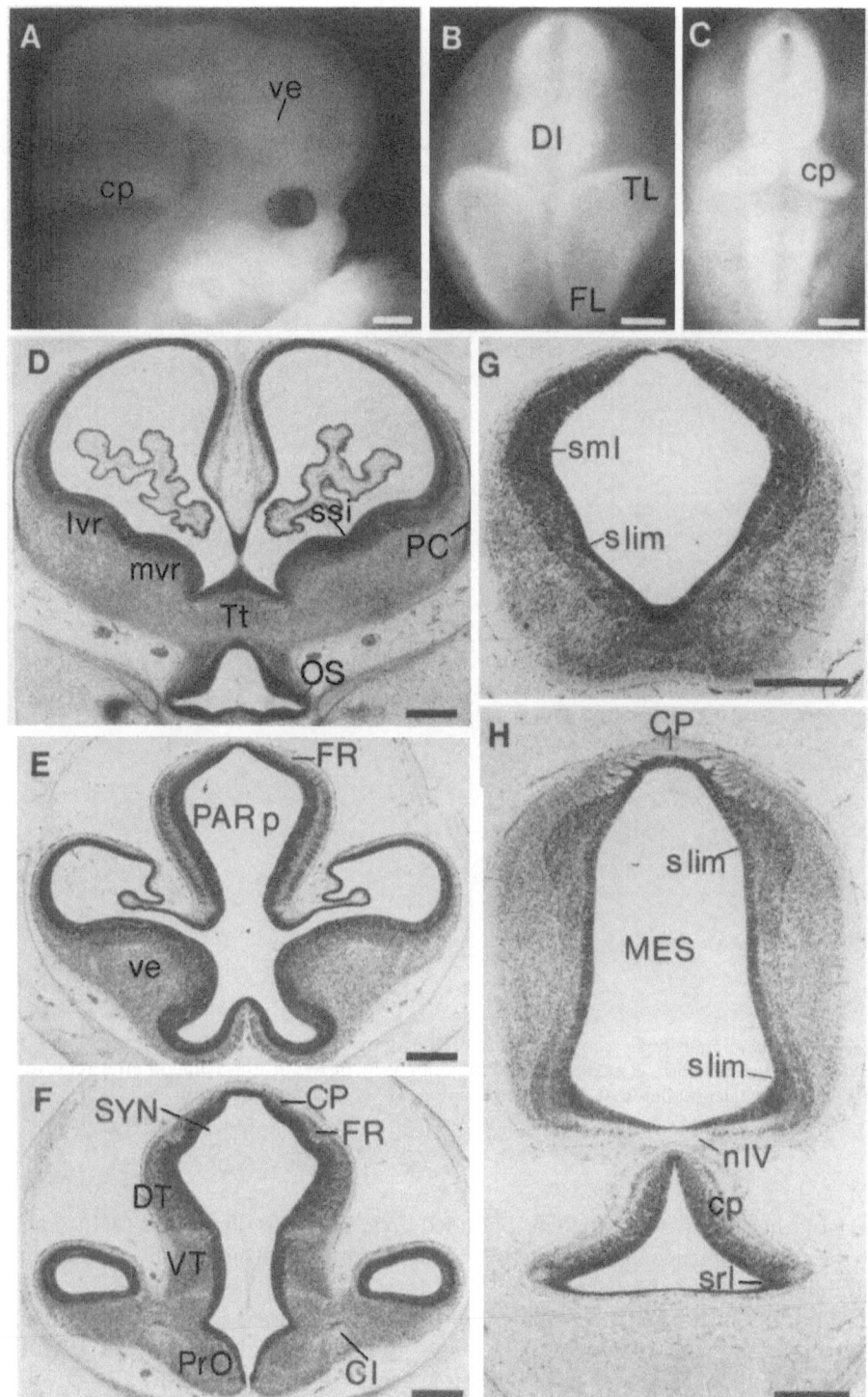

29

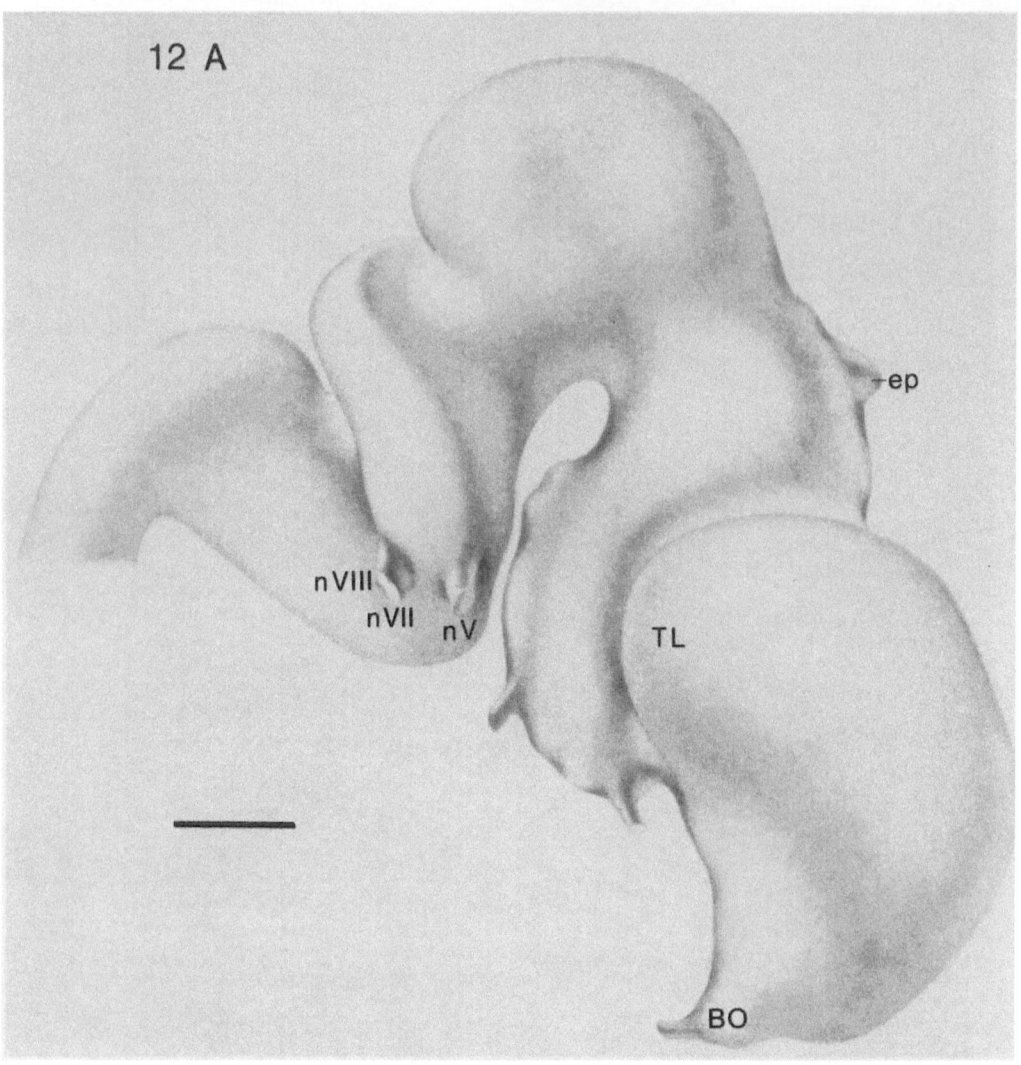

Fig. 12. Three-dimensional reconstructions of the brain of a stage 20 embryo, aged 40 ± 1 days p.c. *A* Lateral view, *B* medial view, *C* superior view of the prosencephalon after removal of parts of the telencephalon and the diencephalon; *calibration bars* = 1.0 mm

part of the alar plate, which is bounded by the sulcus limitans basally and by the sulcus mesencephalicus lateralis dorsally, has also progressively thickened, although less than the basal plate. The dorsal part of the alar plate, located dorsal to the sulcus mesencephalicus lateralis, has preserved a rather thin-walled character. As can be deduced from Figs. 10*B*, 11*G*, and 12*B*, the aspect of four longitudinal zones separated by three ventricular sulci has now also developed in the mesencephalon in contradistinction to the preceding stage. Frontally, the sulcus limitans and the sulcus mesencephalicus lateralis converge towards the synencephalic cavity. Caudally, the sulcus mesencephalicus lateralis ends

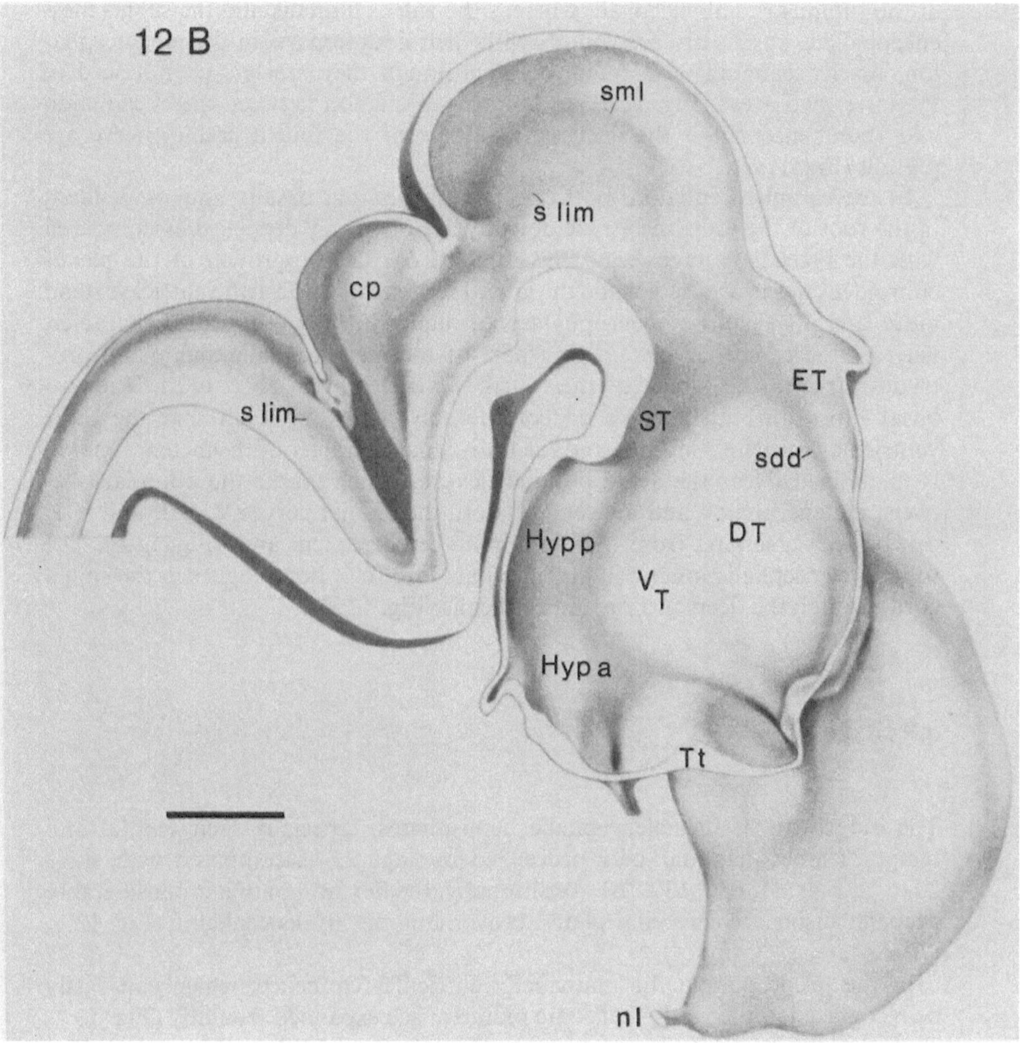

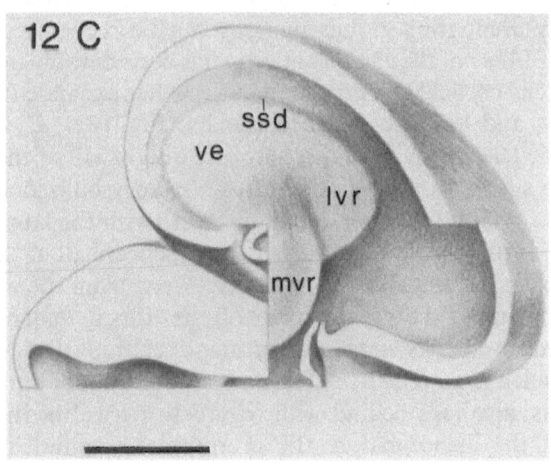

12 B

sml

s`lim

cp

s lim

ET

ST

sdd

Hyp p

DT

V_T

Hyp a

Tt

nl

12 C

ssd

ve

lvr

mvr

Fig. 12 B and C

at the isthmus rhombencephali, whereas the sulcus limitans and the sulcus mesencephalicus basalis are continuous with their counterparts in the metencephalon, myelencephalon, and spinal cord, although they are less pronounced in the latter two areas (Fig. 12 B). In the dorsal junction between mesencephalon and rhombencephalon the decussating fibers of the fourth cranial nerve are present (Fig. 11 H).

In the rhombencephalon, both the flexura pontina basally and its pendant in the roof of the fourth ventricle dorsally are obviously deepened as compared with the preceding stage (e.g., Figs. 10 B, 12 B). The outgrowth of the plexus choroideus has proceeded, while the lateral recesses of the fourth ventricle extend more laterally. The rhombencephalic roof retains its very thin-walled character, whereas the cerebellar plate, being part of the metencephalic alar plate, protrudes far into the lumen of the fourth ventricle (Figs. 11 G, 12 B). The two basal zones have also thickened considerably, with the result that the three ventricular sulci are deepened and the four-zonal character of the metencephalon is underlined. Near the acute pontine flexure, which marks the boundary between metencephalon and myelencephalon, the cranial nerves V, VII, and VIII can be seen emerging from the developing cerebral stem area (Fig. 12 A). The four myelencephalic longitudinal zones pass into their homologues in the spinal cord through the deepened flexura cervicalis (Fig. 12 B).

3.9 Stage 21

The outgrowth of the telencephalic hemispheres, especially their frontal and temporal lobes, has markedly proceeded in stage 21 as compared with stage 20 (e.g., Figs. 11 A, B, 13 A, B). Additionally, the flexura pontina is considerably deeper, causing the cerebellar plates to overhang the myelencephalon (Fig. 13 A, C).

In the telencephalon the undivided ventricular eminence, which can easily be recognized in the methylbenzoate pictures, has expanded frontally (Fig. 13 A, B). Frontally, however, the sulcus subpallii intermedius still separates the medial and lateral ventricular ridges (Fig. 13 D). Within the ventricular ridge area the differentiating globus pallidus can be recognized (Fig. 13 D), and the number of fibers of the capsula interna has considerably increased (Fig. 13 E). The superficial cortical layer (primary cortex) has expanded frontally, dorsally, and laterally, and has also widened basally (Fig. 13 D, E, F). In the frontobasal part of the telencephalic hemisphere, the outgrowth of the olfactory area has proceeded in such a way that a definitive olfactory bulb inclusive of an olfactory ventricle is present, the latter communicating with the lateral ventricle (Fig. 13 E).

In the constituent parts of the diencephalon, the wall-thickening process has extended dorsally and caudally. Apart from the wide hypothalamus and ventral thalamus, the greater part of the dorsal thalamus is now also thick-walled, except for its most dorsocaudal section (Fig. 13 F). The epithalamic region, which is a derivative of the most dorsocaudal part of the parencephalon posterius, also lags behind with regard to its width. In the synencephalic derivatives of the diencephalon, the prerubral tegmental area basally and the pretectal

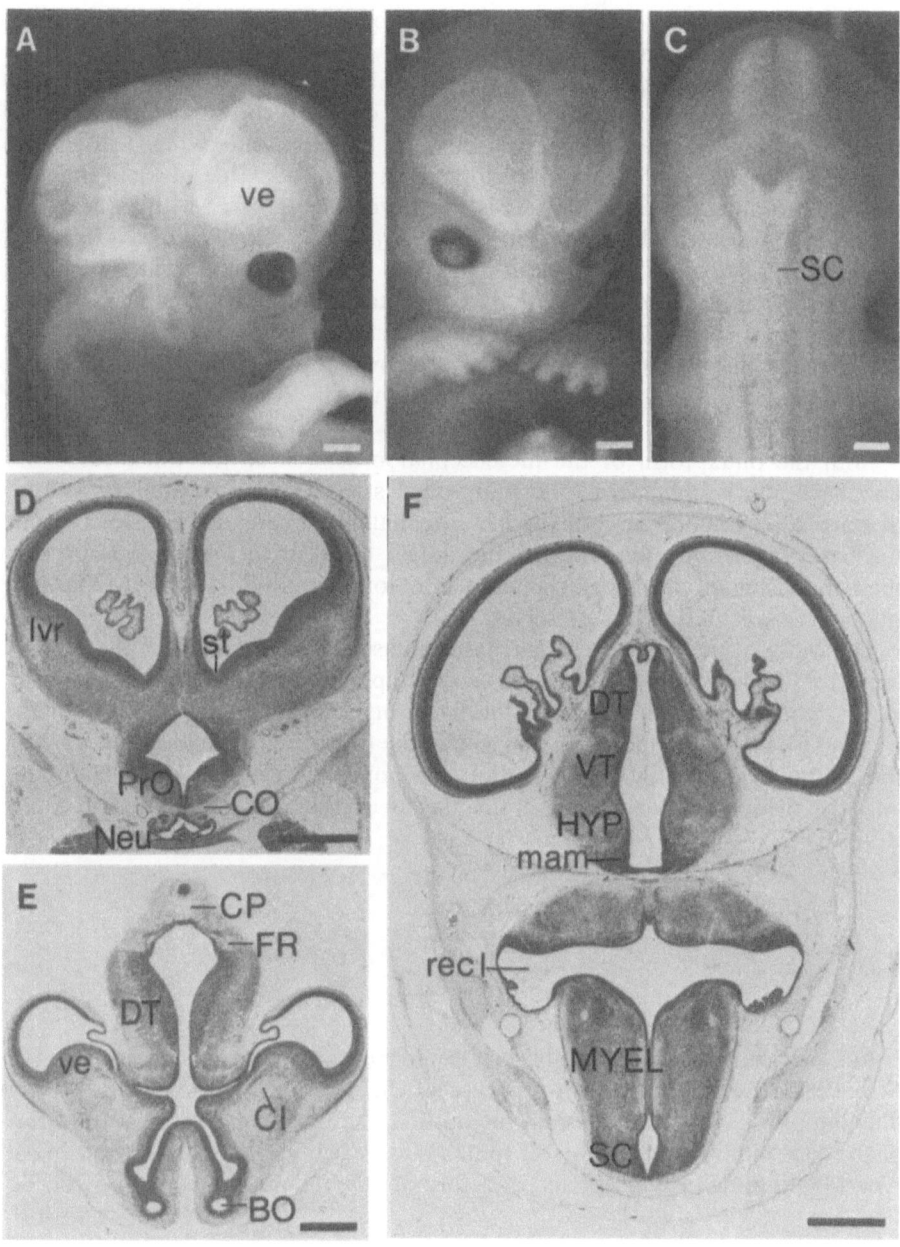

Fig. 13. Photomicrographs of stage 21 embryos, aged 42 ± 1 to 44 ± 1 days p.c. *A, B,* and *C* Lateral, ventral, and dorsal views of the embryo cleared in methylbenzoate. *D* and *F* Horizontal sections, *E* frontal section; *calibration bars* = 1.0 mm

region dorsally, hardly any developmental progress could be detected as compared with the preceding stage.

From the morphogenetic point of view, it is obvious that within the diencephalon as a whole basal sections are ahead of dorsal sections and that caudal parts lag behind frontal parts. It is obvious that the underlying histogenetic process must show two similar developmental gradients. One of them, namely the basal-to-dorsal gradient, is clearly illustrated in Fig. 13F: within the hypothalamus and ventral thalamus initial nuclear anlagen can easily be recognized, whereas in the dorsal thalamus differentiation has not even started. The histogenesis, however, will be dealt with in more detail in a following paper (Gribnau and Geijsberts 1984). Differentiation has also started within the adenohypophysis, whereas the neurohypophysis hardly shows any developmental progress, as can be seen in Fig. 13D, which also shows the broad fiber mass of the chiasma opticum.

Both the outer form of the mesencephalon as well as the thickness of its wall have hardly changed in the embryos of stage 21 as compared with those of stage 20. The thickening of the mesencephalic wall spreads very slowly from the basal tegmental area towards the dorsal tectal area. During this process the subdivision of the mesencephalon into four longitudinal zones separated by three ventricular sulci is preserved.

The width of the fourth ventricle has decreased as compared with the preceding stage, because of the progressive development of the pontine flexure as well as the further thickening of the basal myelencephalon (Fig. 13A, C, F). The outgrowth of the plexus choroideus of the fourth ventricle has now reached the lateral recesses (Fig. 13F). Within the pontine area, the cerebellar plate, and the medulla oblongata several anlagen of cell masses can be observed (Fig. 13F).

3.10 Stage 22

In the pictures of stage 22 embryos cleared in methylbenzoate the outgrowth of both the frontal and temporal lobes of the cerebral hemispheres is clearly illustrated (Fig. 14A, B, C). From the outside, the ventricular eminence in front of the cornu inferius of the lateral ventricle can easily be recognized (Fig. 14A). Apart from these prosencephalic transformations, the outer morphology of the

Fig. 14. Photomicrographs of stage 22 embryos, aged 46 ± 1 to 48 ± 1 days p.c. A, B, and C Lateral, ventral, and superior views of the embryo cleared in methylbenzoate. D, E, and G Horizontal sections; F frontal section; calibration bars = 1.0 mm

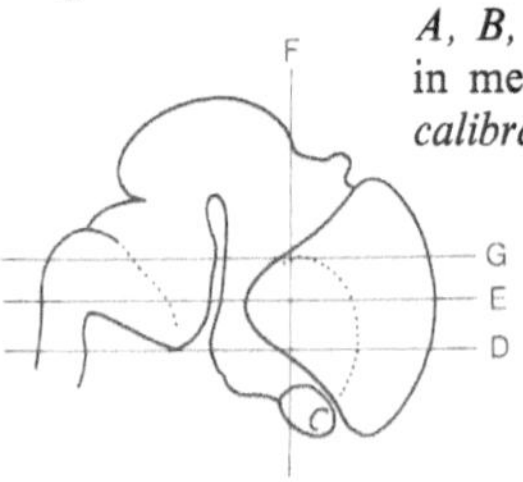

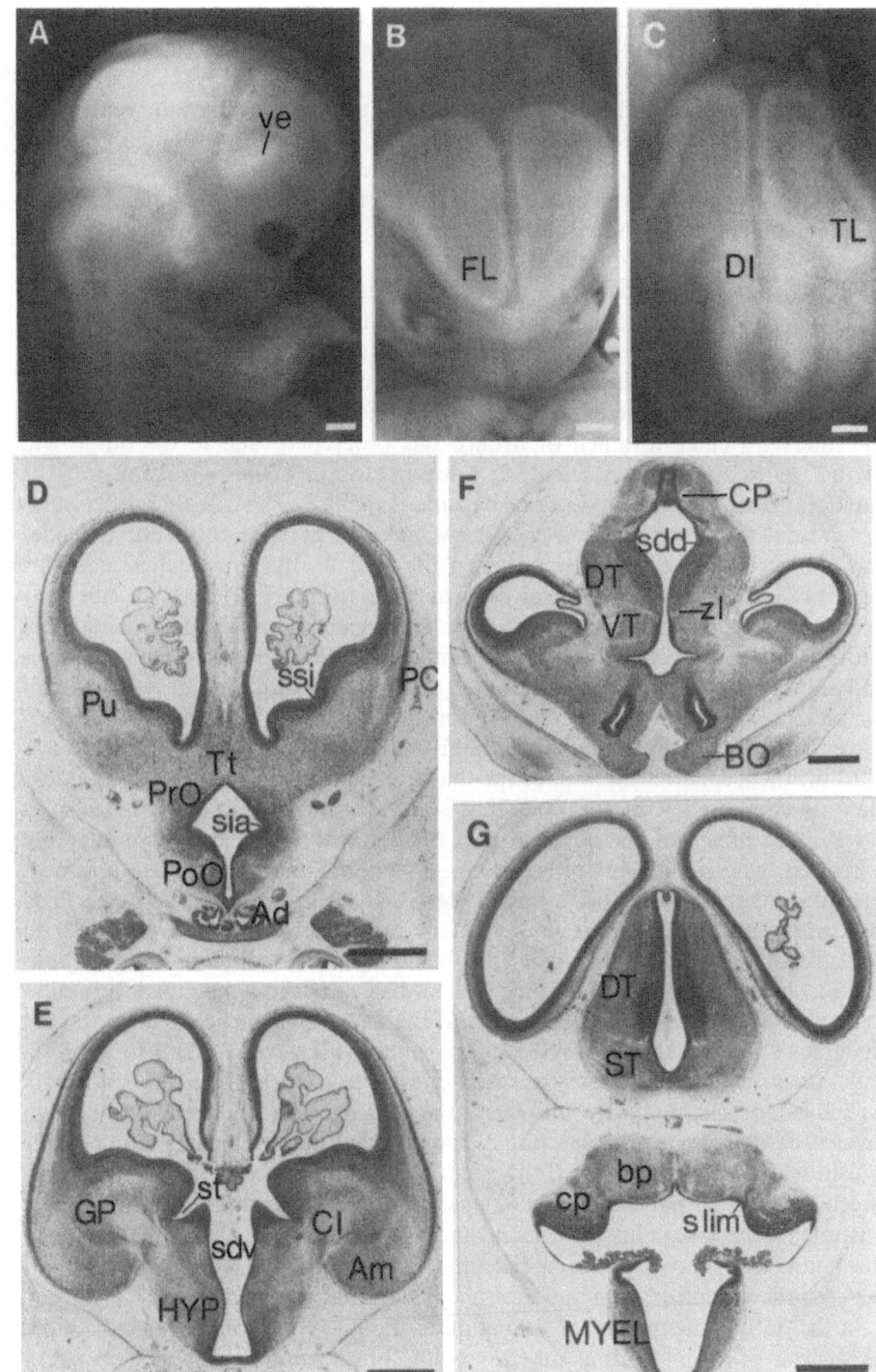

35

brain as a whole has hardly changed as compared with that in the preceding stage (e.g., Figs. 13*A*, 14*A*), probably because of the equal growth of its individual subunits.

Within the telencephalon, the three phases of the generation of the ventricular eminence can still be recognized: (1) frontally, only small parts of the medial and lateral ventricular ridges are present, which are completely separated by the sulcus subpallii intermedius (Fig. 14*D*); (2) in its intermediate part the ventricular eminence shows a shallow sulcus subpallii intermedius, although in its subventricular region the two original ventricular ridges are united (Fig. 14*E*); and (3) caudally, the ventricular ridges as such have vanished completely and a single undivided ventricular eminence, is present (Fig. 14*F*). In comparison with the preceding stage, however, the whole process of unification of the two ventricular ridges has proceeded frontally, in other words, the undivided caudal ventricular eminence has expanded frontally. The narrowing of the interventricular foramen of Monro has proceeded, due to the outgrowth of the ventricular eminence and, to a lesser extent, to the aggregation of the subjacent fiber mass of the internal capsule (Fig. 14*E, F*). A further differentiation within the telencephalic region now makes it possible to identify, besides the globus pallidus, also the putamen, as well as the amygdaloid complex (Fig. 14*D, E, F*). The superficial cortical layer (primary cortex) has not only extended frontally and basally, but has also widened considerably: its maximal width is ten cell layers. The dorsal, primordial pallial part of the hemispheres still shows its thin-walled character (Fig. 14*D, E, F, G*). The development of the olfactory bulb has proceeded considerably (Fig. 14*F*). Its lumen, the olfactory ventricle, is still continuous with the frontal part of the lateral ventricle.

In the diencephalon the wall of the thalamus dorsalis has progressively thickened as compared with the preceding stage (e.g., Figs. 13*E, F,* 14*F, G*). As a result the lumen of the adjacent part of the third ventricle has considerably narrowed, whereas in the epithalamic and hypothalamic regions the lumen of the third ventricle is relatively wide (Fig. 14*E, F*). In the latter area the third ventricle is no longer continuous with the lumen of the neurohypophysis, so the infundibular recess has come into being. As far as differentiation is concerned, the basal parts of the diencephalic wall clearly show the initial formation of individual nuclei, whereas the ventral thalamus and particularly the dorsal thalamus and epithalamus obviously lag behind (Fig. 14*E, F, G*). Morphologically, the ventricular boundary between dorsal and ventral thalamus can hardly be indicated; histologically, however, the demarcation line between the two areas is easily recognizable because of the different developmental state of their constituent walls (Fig. 14*F*). Concurrently, the interneuromeric boundary between the original parencephalon anterius and parencephalon posterius is demonstrated. The fasciculus retroflexus and the posterior commissure demarcate the boundaries between the parencephalon posterius and synencephalon and between the synencephalon and mesencephalon, respectively.

In the developmental stages belonging to the late embryonic phase, the morphogenesis of the individual subunits of the brain caudal to the diencephalon proceeds in such a gradual manner that transformations can hardly be perceived between two successive stages. The morphology of this part of the brain will therefore be described in the section on stage 23, the last embryonic stage to be discussed.

3.11 Stage 23

Developmental stage 23, which was defined in a previous publication as the last stage of the embryonic period, is characterized by the closure of the secondary palate (Gribnau and Geijsberts 1981). The end of stage 23 thus marks the transition from embryo into fetus. In the CNS of embryos belonging to stage 23 the growth of all components has proceeded and remarkable progress has been made in the outgrowth of the constituent parts of the telencephalic hemisphere, as can be seen in the methylbenzoate pictures (e.g., Figs. 14A, B, C, 15A, B, C). The frontal and temporal lobes of the hemisphere, and also its dorsal and occipital parts have expanded in such a way that the external diencephalon is almost completely covered by the telencephalic hemispheres, as is also evident from the three-dimensional reconstruction of the outer aspect of the CNS (Fig. 16A).

In the telencephalon the olfactory bulb is progressively detached from the frontal lobe of the hemisphere (Fig. 16A, B). The lumen of the olfactory ventricle is further reduced, but is still continuous with the lateral ventricle through a narrow channel. Histologically, the olfactory bulb has started to differentiate. The cornu inferius of the lateral ventricle now extends up to a level basolateral to the ventricular eminence (Fig. 15A). The process of merging of the medial and lateral ventricular ridges into one ventricular eminence, which started caudally in stage 19 and proceeded frontally during stages 20, 21, and 22, is now almost complete, except for a small frontal part of the eminence in which a shallow groove can be identified as a remnant of the sulcus subpallii intermedius (Figs. 15D, E, F, 16C, D). Caudobasally, the ventricular eminence is still continuous with the preoptic region via the foramen of Monro, although the lateral extension of the doorstep-like torus transversus marks the boundary between the two areas (Fig. 16B). The torus transversus itself has considerably widened and contains, along with the first fibers of the commissura anterior, the median part of the septal area (Fig. 15D). During later development the corpus callosum will also develop within the torus transversus.

In the pallial area a narrow intermediate zone, consisting of an outer cell-poor layer and an inner cell-rich layer, has developed in between the ventricular zone and the superficial cortical layer or primary cortex (Fig. 15D, E, F, G). The primary cortex itself has expanded dorsally, basally, frontally, and caudally and has also widened: it now consists of up to 25 layers of cells. Nevertheless, the pallial region has retained its thin-walled character as compared with other regions of the brain (Figs. 15D, E, F, G, H, 16B, C, D). The putamen can be easily delimited and the fiber mass of the capsula interna has considerably increased as compared with the preceding stage (e.g., Figs. 14D, E, F, 15D, E, F). Moreover, the globus pallidus and the amygdaloid complex can be identified. As far as the other major fiber systems are concerned, the fornix is already present, in the anterior commissure the first fibers can be recognized crossing the midline within the torus transversus, whereas the corpus callosum is still completely absent (Fig. 15D, E, F, G).

Externally, the changes in the diencephalon are largely confined to the epiphysis, which is now easily recognizable as a separate organ, whereas the neurohypophysis has hardly changed as compared with stage 20 (e.g., Figs. 12A, 16A).

A frontal view of the outer diencephalon is illustrated in Fig. 16 D, which shows the area of both the optic chiasm and the neurohypophysis.

Internally, the morphology of the diencephalon has markedly changed, particularly when a comparison is made between the three-dimensional reconstructions of stages 20 and 23 (e.g., Figs. 12 B, 16 B). The transformations also can be shown by comparing stages 22 and 23 with the aid of the microscopic sections (e.g., Figs. 14 D, E, F, 15 D, E, F, G). The most outstanding morphological feature of stage 23 is the extremely narrowed third ventricle, which in the dorsal thalamic area has even disappeared, resulting in the formation of the adhesio interthalamica (Figs. 15 G, 16 B, C). A second notable feature of the stage 23 diencephalon is the structure of its roof, which has definitively attained the character of a plexus choroideus (Figs. 15 E, F, G, 16 B).

Some of the individual subunits of the diencephalon in stage 23 exhibit little morphogenetic progess compared with stage 20, whereas others have changed markedly (e.g., Figs. 12 B, 16 B). Internally, the telodiencephalic boundary is represented by the curved lateral continuation of the torus transversus towards the foramen of Monro (Fig. 16 B). This ridge also marks the rostral boundary of the preoptic region, which thereby becomes more and more isolated from the ventricular eminence, although originally it is a derivative of the medial ventricular ridge. The torus transversus itself has markedly increased in thickness and has attained a more dorsal position, as has the lamina terminalis. Due to this frontal-to-dorsal shift as compared with stage 20, the preoptic recess, which has considerably widened, now forms the most frontal part of the diencephalon. The sulcus intraencephalicus anterior, curving around the bulge of the preoptic region from the foramen of Monro towards the preoptic recess, clearly marks the caudobasal boundary of the preoptic region as well as the frontal boundary of the hypothalamic region (Fig. 16 B).

The hypothalamic region is bounded by the sulcus diencephalicus ventralis dorsally and by the sulcus diencephalicus basalis ventrally. Caudally, the hypothalamic region exhibits no morphological boundary: it gradually passes into the subthalamic region. Frontally, the hypothalamic region has a broad communication with its counterpart on the opposite side (Fig. 16 B). This communication contains not only the chiasma opticum, but also part of a hypothalamic nuclear anlage which probably eventually develops into the nucleus ventromedialis hypothalami (Fig. 15 D). From the chiasma opticum, the tractus opticus runs dorsally and caudally within the lateral confines of the hypothalamus (Fig. 15 E, F). Caudal to the optic chiasm the recessus postopticus is located (Fig. 16 B). The frontal hypothalamus can be divided into the supraoptic and

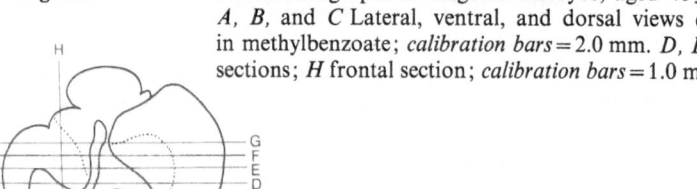

Fig. 15. Photomicrographs of stage 23 embryos, aged 46±1 to 48±1 days p.c. A, B, and C Lateral, ventral, and dorsal views of the embryo cleared in methylbenzoate; *calibration bars* = 2.0 mm. D, E, F, and G Horizontal sections; H frontal section; *calibration bars* = 1.0 mm

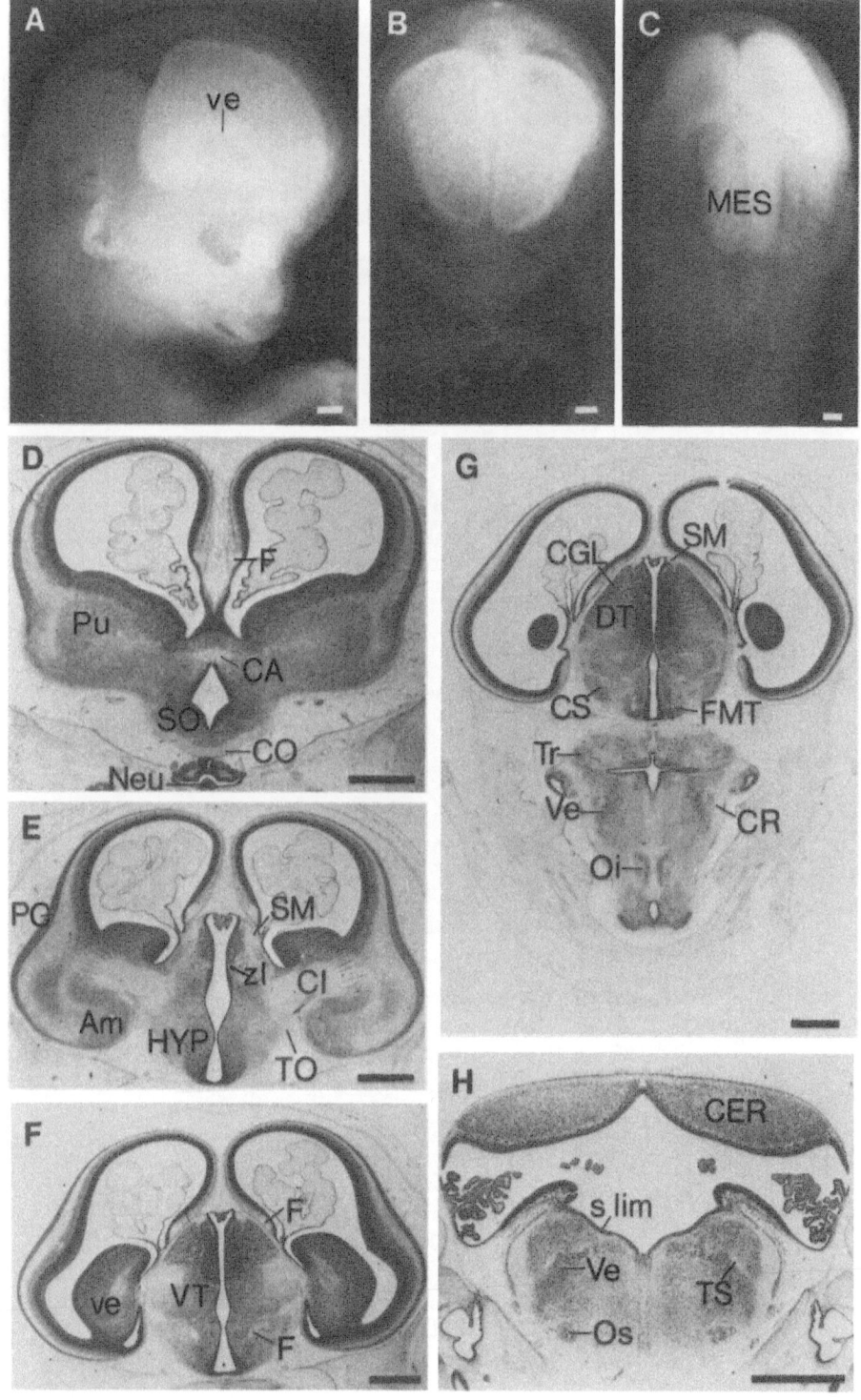

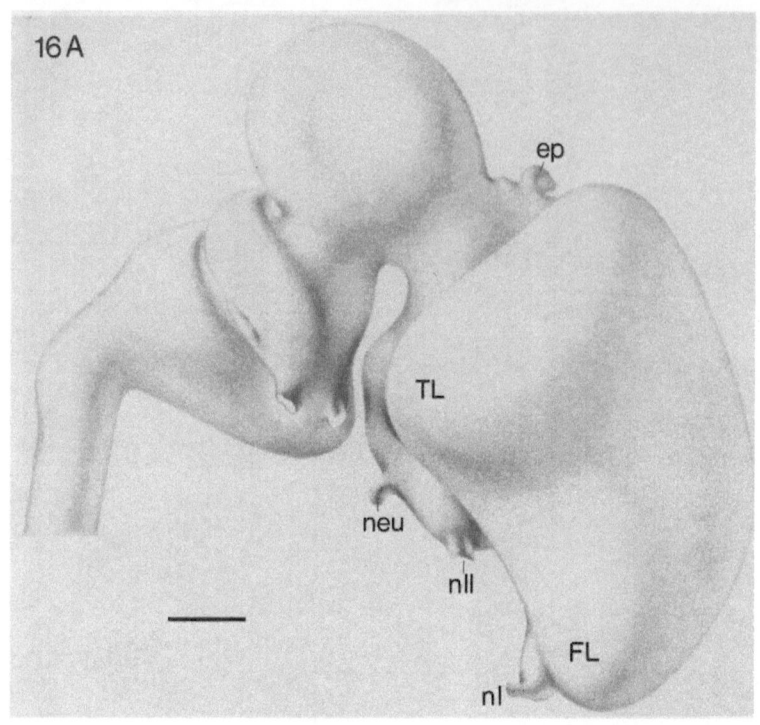

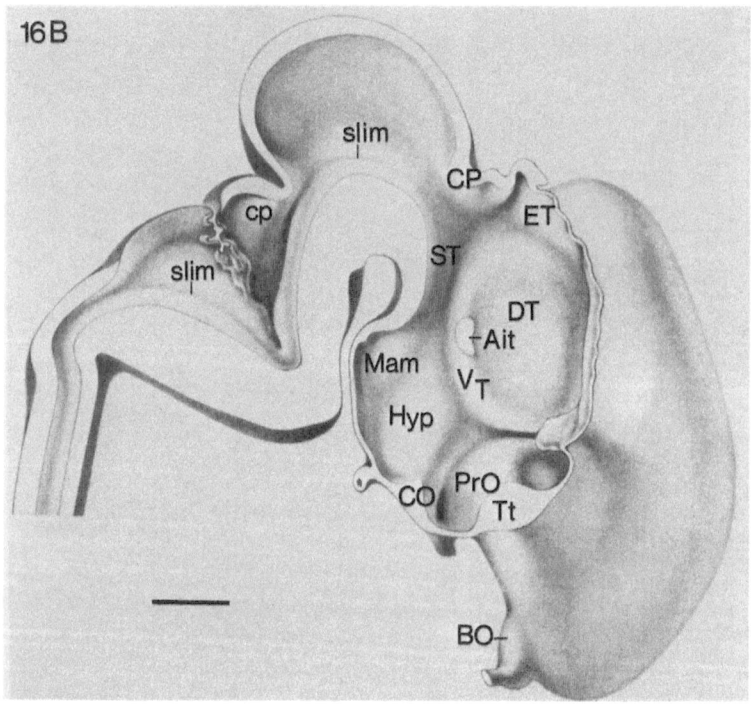

Fig. 16 A and B

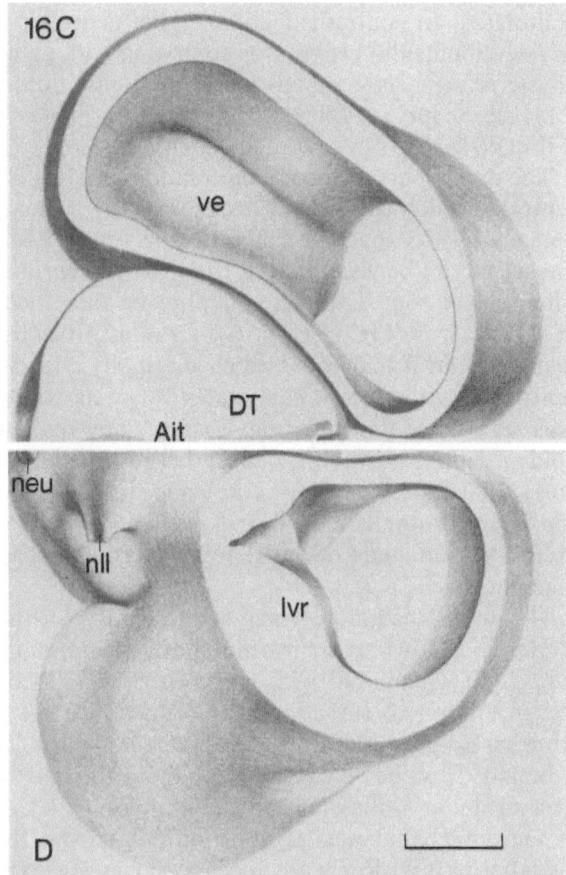

Fig. 16. Three-dimensional reconstructions of the brain of a stage 23 embryo, aged 46 ± 1 days p.c. *A* Lateral view, *B* medial view, *C* superior view of the prosencephalon after removal of parts of the telencephalon and the diencephalon, *D* inferior view of the prosencephalon after removal of parts of the telencephalon; *calibration bars* = 1.0 mm

postoptic regions basally and the anterior hypothalamic region dorsally; morphologically, however, these regions cannot be delimited. The caudal hypothalamus can be divided into the mamillary region basally and the posterior hypothalamic region dorsally, separated by a groove which is a remainder of the sulcus lateralis infundibuli (Fig. 16*B*). In the mamillary region both differentiating mamillary nuclei and developing fiber systems, such as the fornix and the fasciculus mamillotegmentalis, can be identified (Fig. 15*F, G*). In the posterior hypothalamic region the individual nuclear anlagen could not be identified in the present material, but it seems likely that both the lateral hypothalamic area and the posterior hypothalamic nucleus originate within this region.

Besides the hypothalamus, the thin-walled diencephalic floor, which extends from the premamillary recess up to the postoptic recess between the left and right sulcus diencephalicus basalis, is also a derivative of the basal part of the parencephalon anterius (Fig. 16*B*). The floor contains only one characteristic structure, namely the neurohypophysis, the lumen of which is almost completely obliterated. In contradistinction to the neurohypophysis, which shows hardly any developmental progress as compared with preceding stages, the adenohypophysis is in a state of advanced differentiation. Its pars distalis consists of some secondary vesicles with a central lumen and many rosettes of cells (Fig. 15 D).

The dorsal part of the parencephalon anterius is represented by the thalamus ventralis, which is bordered basally by the sulcus diencephalicus ventralis and dorsally by the zona limitans intrathalamica. Morphologically, however, only the sulcus can be visualized in the three-dimensional reconstruction (Fig. 16 B), whereas the zona limitans can only be identified histologically with the aid of the sections (Fig. 15 E, F, G). Graphical reconstruction of the zona reveals that the ventral thalamus, which originally occupied a rather large part of the ventricular surface, now constitutes only a narrow shell around the dorsal thalamus, at least at its ventricular surface. The remainder of the ventral thalamus holds a lateral position, possibly because it is pushed aside by the expanding dorsal thalamus. The ventral thalamic shell extends from the eminentia thalami, which constitutes its most dorsal part and borders upon the foramen of Monro, up to the caudal end of the adhesio interthalamica, which belongs to the dorsal thalamus (Fig. 16 B).

The dorsal thalamus, which is a derivative of the dorsal part of the parencephalon posterius, remains the most outstanding part of the diencephalon (Fig. 16 B). Compared with stage 20 (Fig. 12 B) it is obvious that along with the generation of the adhesio interthalamica the dorsal thalamus also shows a further expansion of its dorsal and caudal parts. Dorsally, the dorsal thalamus is bounded by the sulcus diencephalicus dorsalis, morphologically, and by the stria medullaris, histologically (Fig. 15 E, F, G). The histological appearance of the dorsal thalamus is quite different from that of the ventral thalamus: the latter part is clearly advanced as far as nuclear differentiation is concerned. Within the intermediate layer of the ventral thalamus several nuclear anlagen can be delimited, whereas in the dorsal thalamus only the corpus geniculatum laterale can be identified (Fig. 15 G). This basal-to-dorsal developmental gradient within the diencephalon will be dealt with in more detail in a following paper (Gribnau and Geijsberts 1984).

Dorsocaudally, the sulcus diencephalicus dorsalis deflects ventrally, separating the epithalamic region from the dorsal thalamus; both diencephalic subunits, however, originally derive from the parencephalon posterius (Fig. 16 B). In its roof, the epiphysis cerebri is located, which now shows the compact aspect of a gland with vesicles and follicles present at its periphery, although not shown in Fig. 15. The epithalamic wall has also thickened (e.g., Figs. 12 B, 16 B), and the anlagen of the habenular nuclei can now be identified. From there the fasciculus retroflexus or tractus habenulo-interpeduncularis runs basally, thus marking the transverse boundary between the parencephalon posterius and the synencephalon. At the same time the fasciculus retroflexus caudally delimits the subthalamus, which is the derivative of the basal part of the parencephalon posterius. Dorsally, the subthalamus is bounded by the sulcus diencephalicus ventralis, which caudally fades away into the recessus synencephali (Fig. 16 B). Histologically, the corpus subthalamicum Luysi can be identified in the lateral part of the subthalamus, whereas basally, the fasciculus mamillotegmentalis can be seen passing through towards the mesencephalon (Fig. 15 G).

Morphologically, the synencephalon cannot be delimited, either from the parencephalon posterius or from the mesencephalon (Fig. 16 B). Histologically, the fasciculus retroflexus and the commissura posterior, respectively, mark these boundaries up to the adult stage, as they did in earlier stages (e.g., Figs. 13 E, 14 F). The commissura posterior runs approximately parallel to the fasciculus retroflexus in a transverse plane through the sulcus mesodiencephalicus. The basal part of the synencephalon is rather thick-walled, whereas its dorsal part is relatively thin-walled, surrounding the recessus synencephali (Fig. 16 B). As far as nuclear differentiation is concerned, the identification of individual nuclei is hardly possible, either in the basal or in the dorsal part, which constitute the prerubral tegmental and pretectal areas, respectively. During further development the former area will probably give rise to the nucleus interstitialis (Cajal) and the nucleus of Darkschewitsch, whereas in the latter area the pretectal nuclei and the nucleus posterior thalami will arise (e.g., Keyser 1972). The basal synencephalic area is pierced by the fasciculus mamillotegmentalis, as is the basal parencephalon posterius.

Externally, the mesencephalon has not changed very much: it retains the appearance of a single bulge, of which the lateral walls seem to be flattened (Fig. 16 A). Internally, the morphology of the mesencephalon in stage 23 shows some slight transformations as compared with that in stage 20 (e.g., Figs. 12 B, 16 B). The four-zone pattern is replaced by a three-zone pattern, because the sulcus mesencephalicus basalis has vanished while the sulcus limitans and the sulcus mesencephalicus lateralis are still present. As a consequence, the two original longitudinal zones of the basal plate now form one single zone which is dorsally bounded by the sulcus limitans. The latter sulcus constitutes the borderline with the alar plate, which in its turn is divided into two longitudinal zones by the sulcus mesencephalicus lateralis (Fig. 16 B). Rostrally, both sulci fade away in the recessus synencephali, caudally, however, the sulcus mesencephalicus lateralis ends in a caudolateral extension of the mesencephalic bulge, whereas the sulcus limitans passes beyond the isthmus into the pontine and medullary regions. Histologically, the wide basal tegmental area is further advanced than the thin-walled tectal area. Within the former region the pedunculus cerebri can be recognized, as well as the primordial nucleus ruber, substantia nigra, and nucleus interpeduncularis, whereas the fasciculus retroflexus and the fasciculus mamillotegmentalis both end in this area. In the tectal region, however, the intermediate layer consists of widely spread cells in which differentiation has not yet begun.

Externally, the rhombencephalon is divided into its component parts, the metencephalon and the myelencephalon, by a deep indentation of its extremely thin roof at the level of the flexura pontina (Fig. 16 A). Within the metencephalon an external groove separates its basal part or pontine area from its dorsal part, which corresponds to the rhombic lip. Basally, the cranial nerves V, VII, and VIII can be seen arising from the brain stem near the pontine flexure, whereas the remaining cranial nerves IX–XII are not depicted in Fig. 16 A.

The internal morphology of the rhombencephalon in stage 23 shows a high consistency with that in stage 20 (e.g., Figs. 12 B, 16 B). The plexus choroideus of the fourth ventricle has extended laterally up to the lateral recesses (Fig. 15 H). In the metencephalon, the frontal part of the alar plate, containing the cerebellar anlage, has progressively widened and is fused with its counterpart on the opposite side, forming a broad median communication. Moreover, this widening is coupled with a deepening of the sulcus limitans, which delimits it from the metencephalic basal plate (Fig. 16 B). The basal sulcus rhombencephalicus has vanished, leaving the brain stem area with a three-zone pattern, with the median sulcus rhomboideus inferior, the sulcus limitans, and the sulcus rhombencephalicus lateralis, in basal-to-dorsal order (Fig. 15 H). Histogenetically, the differentiation of brain stem nuclei is clearly advanced as compared with nuclei in other regions of the central nervous system: the nuclei of the cranial nerves as well as the olivary nuclei could be identified with certainty (Fig. 15 G, H). Of the fiber systems, both the corpus restiforme and the tractus solitarius are shown in the sections, whereas the tractus pyramidalis could not be delimited from the surrounding white matter.

In the spinal cord, the alar plate has thickened considerably in comparison with stage 20 (e.g., Figs. 12 B, 16 B). The morphological pattern of four longitudinal zones separated by three ventricular sulci is maintained.

4 Discussion

In the previous chapter a comprehensive description of the morphogenesis of the CNS of the rhesus monkey has been presented. In the ensuing chapter our findings will be placed into a comparative embryological perspective. Attention will be focussed on the following aspects: (1) neuromerism, (2) longitudinal zones and ventricular sulci, (3) the morphological plan of the diencephalon, (4) the basal forebrain, and (5) the chronology of the morphogenesis of the brain.

4.1 Neuromerism

As was pointed out in the Introduction, the mere presence of neuromeres in the developing vertebrate brain both before and after closure of the neural tube is generally accepted in literature. The number of neuromeres, however, is still a matter of dispute as reviewed by Vaage (1969) and Keyser (1972). Vaage stated that the varying numbers of neuromeres suggested by various authors can readily be explained by the fact that the number depends on the developmental stage. In studying closely graded stages in the chick, Vaage (1969) concluded the existence of maximally seven rhombomeres, two isthmic neuromeres, two mesencephalic neuromeres, and eight prosencephalic neuromeres, six of the latter being diencephalic and two telencephalic. Keyser (1972), on the other hand, discerned only five neuromeres in the hamster diencephalon and one in the telencephalon, since he did not subdivide the synencephalon or the telencephalon into two separate neuromeres.

In the rhesus monkey neuromeres can be clearly distinguished (see Fig. 17). In stage 13 embryos, seven rhombomeres and two isthmic neuromeres could be identified, which remain distinct up to stage 15 inclusively. Although much less conspicuous they are still present during stage 16, but in stage 17 they have completely vanished. The otic vesicle is located opposite rhombomere five. Similar results were described in primate embryos by Müller and O'Rahilly (1980). During stage 15 two individual neuromeres could be discerned within the isthmus region. Thus, as far as the rhombencephalon is concerned, the neuromeric pattern of the rhesus monkey closely resembles that of the chick as described by Vaage (1969).

In the rhesus monkey mesencephalon two neuromeres, m_1 and m_2, could be identified during developmental stages 14 through 17. These two mesencephalic neuromeres are much less distinct than the rhombomeres. The presence of two transient neuromeres during early development was also noted by Vaage (1969) in the chick and by Keyser (1972) in the Chinese hamster.

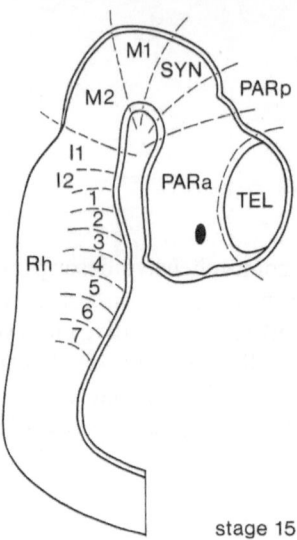

stage 15

Fig. 17. Neuromery in the rhesus monkey brain, schematically shown in stage 15. A total of seven rhombomeric, two isthmic, two mesencephalic, three diencephalic, and one telencephalic neuromeres were found

In the diencephalon of the rhesus monkey we had to distinguish between the caudal and rostral parts. The two most caudal diencephalic neuromeres, the synencephalon and the parencephalon posterius, could easily be identified during stage 13 and stage 14. In stages 15 and 16, besides these two neuromeres the dorsal part of the parencephalon anterius could also be recognized. During stage 17, however, the latter neuromere vanished, whereas both the synencephalon and the parencephalon posterius remain visible up to stage 20 inclusively. In stage 19 the thickening of the wall of the parencephalon posterius starts to supersede the neuromeric pattern. In stage 21 the diencephalic neuromeres have completely vanished although the neuromeric borders between the two parencephalic neuromeres, the parencephalon posterius and the synencephalon, and caudal to the synencephalon, remain recognizable by the presence of the zona limitans intrathalamica, the fasciculus retroflexus, and the commissura posterior, respectively. The neuromeric pattern of the caudal diencephalon is in accordance with that in the Chinese hamster as described by Keyser (1972), in which, in contradistinction to that of the chick (Vaage 1969), only a single synencephalic neuromere is present.

In the literature, quite different opinions exist about the neuromerism in the frontal part of the diencephalon. Some authors (Bergquist 1952a, b; Bergquist and Källén 1954, 1955) describe just one optic neuromere in front of the parencephalon anterius, whereas others discern two frontal diencephalic neuromeres: an optic and a postoptic one (Vaage 1969; Keyser 1972). In our material from the rhesus monkey no neuromeres could be identified in the frontal diencephalon: the parencephalon anterius grades into the optic region without any boundary. Two possible explanations can be given for this finding: either there is just one frontal neuromere, as suggested by the Swedish authors,

46

or there are indeed two neuromeres present in earlier stages of development, which have vanished at stage 13 in the rhesus monkey. We were not able to decide for one out of these two possibilities, also because neuromerism in the frontal part of the diencephalon culminates in very early stages. For more detailed information the reader is referred to Sect. 4.3.

4.2 Longitudinal Zones and Ventricular Sulci

The morphogenesis of the CNS cannot be understood without a basic knowledge of the early histogenetic events, which will be dealt in more detail in a following paper (Gribnau and Geijsberts 1984). The first phase of the histogenesis of the brain is characterized by cell production, which initially results in longitudinal growth of the CNS. During further development, however, continuous cell production is gradually accompanied by radial migration of an increasing number of cells. Consequently, longitudinal growth is then attended by radial growth. The latter is expressed in an increasing width of the cerebral wall, together with a decreasing ventricular lumen. The histogenetic events described above are characterized by a heterochronous course: they proceed at different times and with different rates in the individual parts of the CNS. In addition, the brain shows both a basal-to-dorsal and a caudal-to-rostral developmental gradient. As a consequence, from two adjacent areas of the brain either one may thicken and the other not, or they will thicken one after the other resulting in a temporary or permanent segregation by a groove. In this way the histogenetic process of the brain has a considerable impact on its morphogenesis.

In the rhesus monkey, the basal part of the rhombencephalic wall starts to thicken during stage 14: a vague sulcus limitans can be identified. In stage 15 the rhombencephalon shows, apart from the neuromeric pattern in its basal part, a longitudinal zonal pattern: the basal plate, subdivided by the sulcus rhombencephalicus basalis into a ventral and a dorsal zone, the latter inclusive of its rhombomeres; then the sulcus limitans and the alar plate. In the rostral part of the alar plate, in which the cerebellum will originate, a third sulcus, the sulcus rhombencephalicus lateralis, is added. Thus, here a longitudinal zonal pattern with four zones is present: the alar plate is divided into a dorsal and a ventral zone. During further development the four-zonal pattern also arises in the myelencephalic part of the rhombencephalon. In stage 17, the rhombomeres have completely vanished and the entire rhombencephalon at that time exhibits the pattern of four longitudinal zones, congruent with the His model: the somatomotor, the visceromotor, the viscerosensoric, and the somatosensoric zones.

In the mesencephalon the pattern of four longitudinal zones arises in a similar way but somewhat later. In stage 15 two longitudinal sulci are present in the mesencephalon: the sulcus limitans and the sulcus basalis mesencephali. Thus a pattern is established of three longitudinal zones separated by two ventricular sulci: the basal plate, consisting of two longitudinal zones separated by the sulcus mesencephalicus basalis, is delimited from the alar plate by the sulcus limitans. This pattern is retained up to stage 20 in which a third groove, the sulcus lateralis mesencephali, is added within the alar plate, and thus the pattern of four longitudinal zones is achieved. The progressive outgrowth of the two basal zones as well as the fusion of the basal parts of both sides results in the disappearance of the sulcus mesencephalicus basalis and the formation of the undivided tegmental area in stage 23. In that stage the tectal area of both sides is still divided into two longitudinal zones by the sulcus mesencephalicus lateralis. Rostrally, however, the mesencephalic sulci fade away into the synencephalic recess and thus their continuation into the sulci of the diencephalon could not be determined.

In the diencephalon, the presence of two vague horizontal sulci in the basal parencephalon anterius during stage 13 suggests an early longitudinal zonation in the frontal part of the diencephalon. During stage 14 this pattern becomes more evident and in stage 15 it is accentuated by the presence of two longitudinal grooves: the sulcus diencephalicus basalis and the sulcus diencephalicus ventralis. The hypothalamic cell cord, which develops in between these two sulci, is the first longitudinal zone present in the diencephalon and becomes more prominent during stage 16. In stage 17 a second longitudinal zone is added in the frontal diencephalon, located dorsal to the sulcus diencephalicus ventralis. This zone consists of the preoptic region, which is a derivative of the medial ventricular ridge, rostrally, and the anlage of the ventral thalamus, which is a derivative of the dorsal part of the parencephalon anterius, caudally. The two are separated by the sulcus intraencephalicus anterior.

In the caudal diencephalon, the thin-walled neuromeric character is preserved up to stage 17. In that stage an incipient thickening of the basal parts of both the parencephalon posterius and the synencephalon could be noted. Although during stage 18 the widening of these basal parts gradually advances, only in stage 19 does the caudal diencephalon clearly get involved in the longitudinal zonation process inclusive of the formation of ventricular sulci. In that stage the thickening of the basal parts of the parencephalon posterius and the synencephalon has proceeded, representing the subthalamic and prerubral tegmental area respectively. Dorsally, the rostral part of the parencephalon posterius starts to thicken in stage 19; it is dorsally bounded by the curved, shallow sulcus diencephalicus dorsalis. The remaining dorsocaudal part of the parencephalon posterius is still thin-walled as is the dorsal synencephalon.

In stage 20 the morphology of the entire diencephalon is characterized by more or less thick-walled longitudinal zones separated by ventricular sulci. The diencephalic floor, which is dorsally bounded by the sulcus diencephalicus basalis, has retained its rather thin-walled character. The basal diencephalon, consisting of the derivatives of the hypothalamic cell cord as represented by the postoptic and anterior hypothalamic regions anteriorly, is a rather thick-walled longitudinal zone, dorsally bounded by the sulcus diencephalicus ventralis. Caudally, this zone passes into the subthalamic region and thereupon into the prerubral tegmental area. The dorsal diencephalic longitudinal zone is composed of, in frontal-to-caudal order: the preoptic region, the ventral thalamus, the dorsal thalamus, the epithalamus, and the pretectal area.

The latter two regions are still relatively thin-walled and therefore a groove separating this area from the thin-walled diencephalic roof is absent. Rostrally, however, the wide thalamus is dorsally bounded by a deep sulcus diencephalicus dorsalis, which delimits the dorsal diencephalic zone from the thin diencephalic roof.

During further development the longitudinal zonal pattern of the diencephalon is accentuated by an advanced widening of the individual subunits, especially in the caudodorsal zone represented by the dorsal thalamus, the epithalamus, and the pretectal region. In this way, the morphological pattern of the diencephalon is consistent with that in the brain stem: besides the floor plate and the roof plate, there are two longitudinal zones bounded by three ventricular sulci, the latter represented by the basal, ventral, and dorsal diencephalic sulci. However, the diencephalic longitudinal zones cannot be homologized with the longitudinal zones of the brain stem.

In conclusion it can be stated that the longitudinal zones result from the histogenetic process occurring within the wall of the CNS. Prior to the onset of that process, however, the CNS is divided into transversely oriented neuromeres, which gradually become transformed into longitudinal zones. This transformation is accomplished via both caudal-to-rostral and basal-to-dorsal developmental gradients. In our opinion, the longitudinal zonal pattern is superimposed upon the transverse neuromeric units by the chronologically varying histogenetic activity of their respective basal and dorsal parts. Thus the fundamental morphological subunits of the brain are the primary neuromeres, rather than the secondary longitudinal zones. Keyser (1972), in his extensive study on the development of the diencephalon of the Chinese hamster, came to the same conclusion.

4.3 The Morphological Plan of the Diencephalon

Several authors have been engaged in unraveling the ontogenesis of the prosencephalon in various species, as was reviewed by Keyser (1972). In view of the implications on the morphological plan of the diencephalon, the most outstanding research was accomplished by Bergquist and Källén (1954, 1955), Coggeshall (1964), Vaage (1969), and Keyser (1972). Our objective was to analyze whether the conclusions reached by these authors in various lower vertebrates, birds, and rodents also apply to primates. For a better understanding, their findings are schematically transformed upon our stage 19 prosencephalon and, together with our results, represented in Fig. 18.

On the basis of their comparative studies on various lower vertebrates, Bergquist and Källén (1954) concluded the existence of one telencephalic and three diencephalic neuromeres. In a subsequent experimental study on *Ambystoma,* however, the same authors (1955) stated that the appearance of both the optic and telencephalic neuromeres was merely a secondary phenomenon resulting from lateral evaginations, and not neuromeres in the real sense. They thus divided the prosencephalon into two neuromeres, a and b, in which b is identical with our synencephalon, whereas a comprises the rest of the prosencephalon, a_3, with two bilateral evaginations: an optic one, a_2, and a telencephalic one, a_1 (Fig. 18).

In the rat prosencephalon, Coggeshall (1964) identified three neuromeres: an anterior, a middle, and a posterior one; I, II, and III, respectively (Fig. 18). His neuromeres III and II are identical with our synencephalon and parencephalon posterius respectively, whereas his neuromere I consists of both our parencephalon anterius and our telencephalic neuromere.

Quite another opinion was held by Vaage (1969) who in the chick prosencephalon discerned six diencephalic neuromeres (n_3-n_8), in which n_7 and n_8 represent two synencephalic neuromeres and n_3 and n_4 the optic and postoptic neuromeres. Additionally, this author distinguished two telencephalic neuromeres, n_1 and n_2 (Fig. 18). The neuromeric pattern of the Chinese hamster prosencephalon, as suggested by Keyser (1972) is similar to that of Vaage, except for the presence of only one synencephalic as well as only one telencephalic neuromere. Thus Keyser discerned a total of six prosencephalic neuromeres (I–VI): the synencephalon, parencephalon posterius, parencephalon anterius, and the postoptic, optic, and telencephalic neuromeres (Fig. 18).

On the basis of our rhesus monkey material we concluded the existence of one synencephalic neuromere, which is in accordance with the results of all other authors except for Vaage. Anterior to the synencephalon, we were able to identify the parencephalon posterius corresponding with that of Keyser (1972) and Vaage (1969) and the middle neuromere V of Coggeshall (1964). In contradistinction to both Keyser and Vaage, however, the anterior part of the diencephalon in our opinion consists of one single neuromere, the parencephalon anterius, having a bilateral optic evagination. As opposed to Bergquist and Källén (1955), Coggeshall (1964), and Vaage (1969), however, we concluded the presence of a single telencephalic neuromere, as was also suggested by Keyser (1972).

During further development, the neuromeric pattern of the diencephalon is gradually transformed into a longitudinal zonal pattern, as described in the previous section. Subsequently, the basal and dorsal diencephalic longitudinal zones, which are bounded by the sulcus diencephalicus basalis, ventralis, and dorsalis, respectively, become distorted because of a differential growth of their respective subunits, the dorsal thalamus in particular (Fig. 19). The original interneuromeric borders, however, remain traceable because of the presence of developing fiber systems. The syn-mesencephalic boundary is characterized by the posterior commissure, the syn-parencephalic boundary by the fasciculus retroflexus, and the interparencephalic boundary by the zona limitans intrathalamica.

In conclusion it can be stated that the morphogenetic plan of the diencephalon in the rhesus monkey is represented by three successive neuromeres which become divided by the sulcus diencephalicus ventralis into a basal zone and a dorsal zone. Eventually, the individual parts of these zones will develop into the respective diencephalic regions of the adult brain, as summarized in Table 2. There is one notable difference between our results and those of Keyser (1972) in the Chinese hamster, namely the origin of the preoptic region. In Keyser's view, the preoptic region arises as the most anterior part of the hypothalamus and thus would be a derivative of the basal zone. According to that author, the rostral ending of the sulcus diencephalicus ventralis, which marks the bound-

Fig. 18. Schematic representation of the neuromeric pattern of the prosencephalon as suggested ▷ by various authors and transferred to the stage 19 rhesus monkey brain; for a discussion, see text

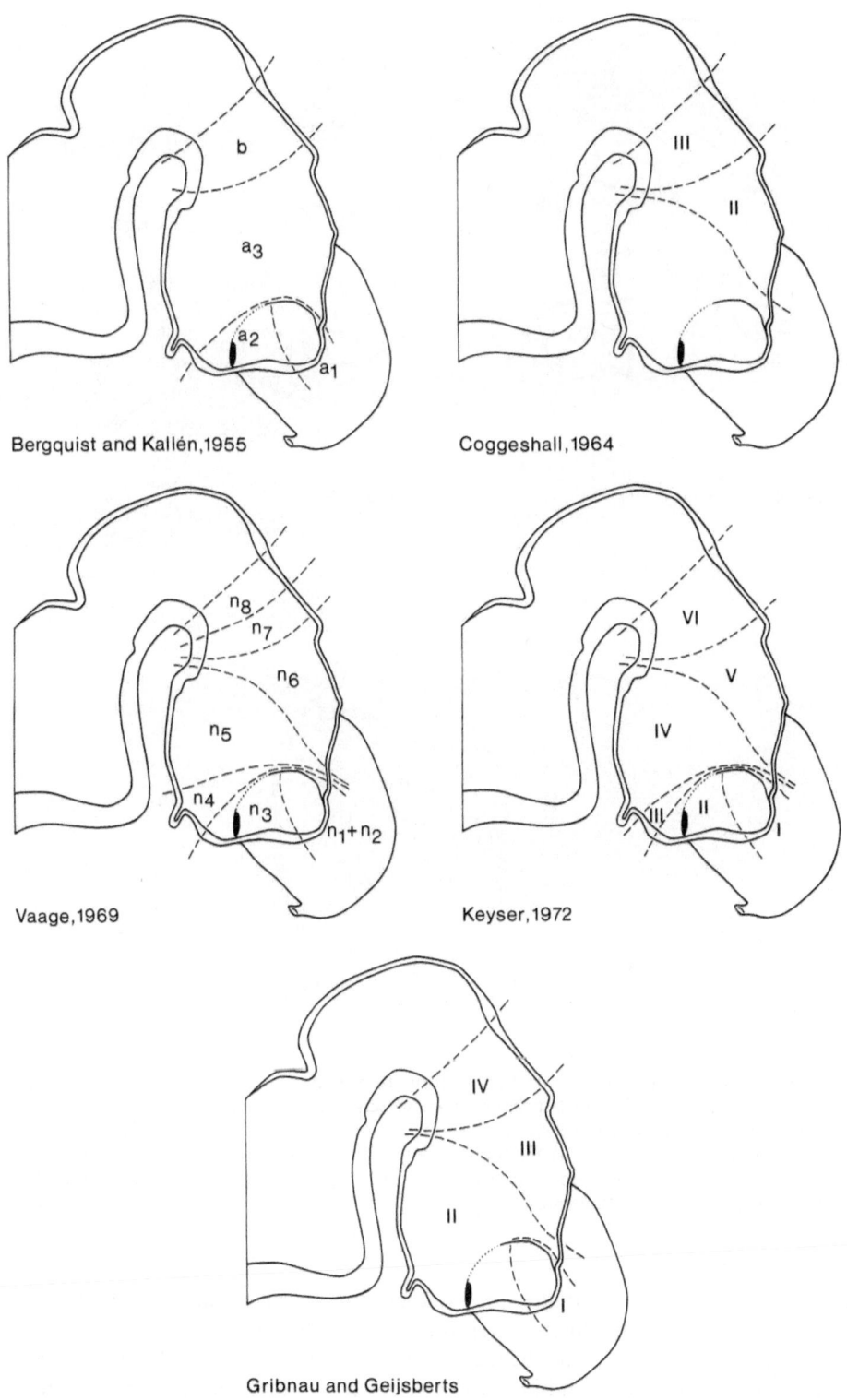

Bergquist and Kallén,1955

Coggeshall,1964

Vaage,1969

Keyser,1972

Gribnau and Geijsberts

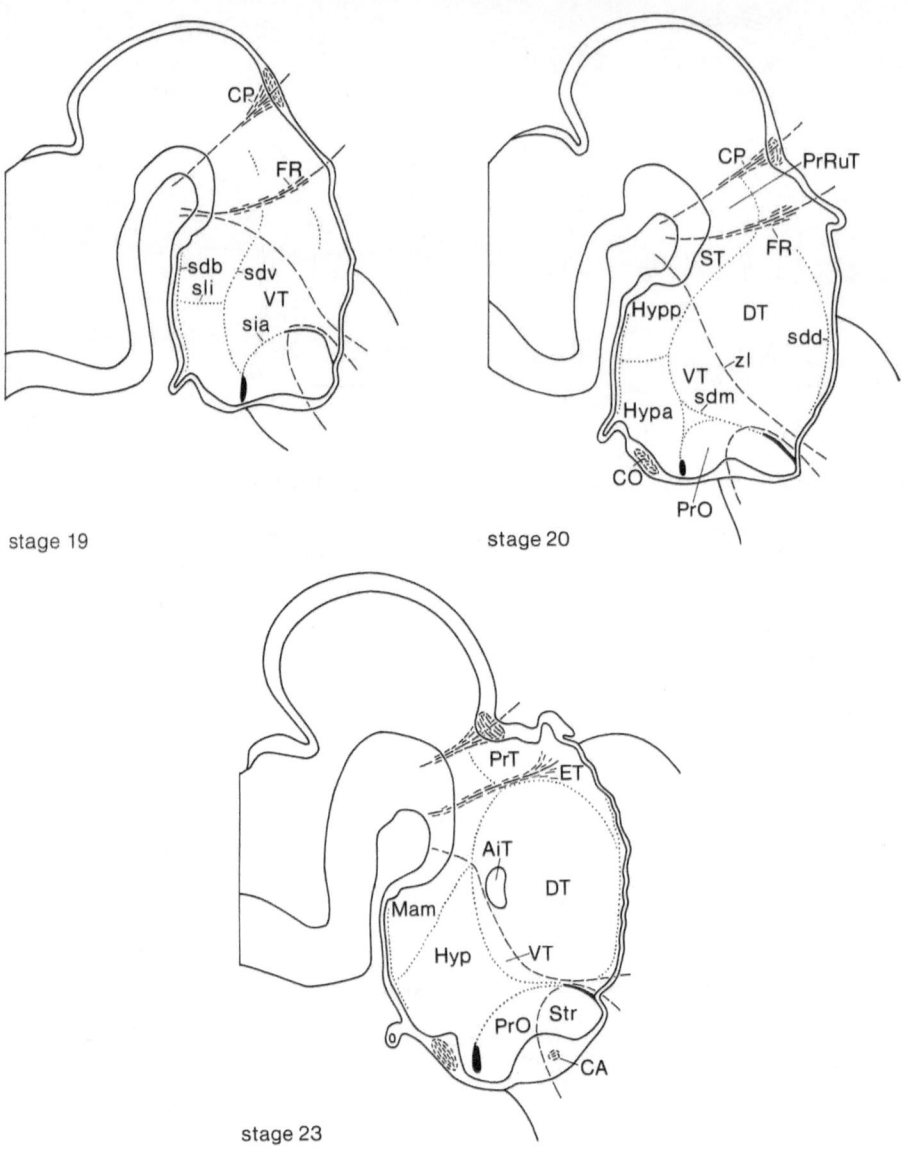

stage 19

stage 20

stage 23

Fig. 19. Schematic representation of the transformations taking place in the developing rhesus monkey diencephalon. The three primary, transversely oriented neuromeres are superseded by longitudinal zones, which in their turn become distorted, mainly because of the disproportionate outgrowth of the dorsal thalamus. The original interneuromeric boundaries, however, remain visible because of the presence of the commissura posterior, the fasciculus retroflexus, and the zona limitans intrathalamica, respectively

ary between the basal and the dorsal zone, during early stages is situated in the preoptic recess but during later stages in the foramen of Monro. In our opinion, however, the sulcus diencephalicus ventralis ends rostrally in the preoptic recess both in early and in late developmental stages and thus the preoptic region originates in the most anterior part of the dorsal zone of the parencepha-

Table 2. Ontogenetic units of the diencephalon

Neuromere	Basal zone	Dorsal zone
— — — — — — — — Crista telodiencephalica— —		
	Supraoptic region	
	Postoptic region	Preoptic region
Parencephalon anterius	Anterior hypothalamic region – Sulcus intraencephalicus anterior	
	— — Sulcus lateralis infundibuli — — —	
	Posterior hypothalamic region	Ventral thalamus
	Mamillary region	
— — — — — — — — Crista interparencephalica — — — — — — Zona limitans intrathalamica — — —		
Parencephalon posterius	Subthalamic region	Dorsal thalamus
		Epithalamus
— — — — — — — — Crista synparencephalica — — — — — — — Fasciculus retroflexus — — — — — —		
Synencephalon	Prerubral tegmentum	Pretectal area
— — — — — — — — Crista synmesencephalica— — — — — — — Commissura posterior— — — — — —		

lon anterius. Probably the difference is related to the fact that we had three-dimensional reconstructions at our disposal instead of graphical reconstructions like those used by Keyser.

4.4 The Basal Forebrain

The morphogenesis of the basal forebrain is dominated by two outstanding features: (1) the evagination of the telencephalic hemispheres and, connected with that, the demarcation of the telodiencephalic boundary, and (2) the arisal of two ventricular ridges in each hemisphere and their merging into one ventricular eminence. Coherent with the latter process is the blurring of the basal part of the telodiencephalic boundary and the origin of the preoptic region.

4.4.1 The Evagination of the Telencephalic Hemispheres

The prosencephalon of stage 13 rhesus monkey embryos consists of a narrow posterior part and a wide anterior part of the neural tube. During further development the former will be transformed into the caudal part of the diencephalon, whereas the latter will give rise to both the anterior part of the diencephalon and the telencephalon. The wide anterior part apparently originated by a dorsally directed evagination of the future telencephalic area of the neural tube; a telodiencephalic boundary, however, is absent at that time. In the present investigation, the telo-diencephalic boundary is defined as the velum transversum–torus transversus line after Keyser (1972), who gave an extensive review of the literature available on this matter.

During stage 14 a bilateral evagination is added to the dorsal expansion of the telencephalic area, especially in its caudal part. As a consequence the telodiencephalic boundary becomes visible within its dorsal-to-caudal range, where a sulcus telodiencephalicus is developing on the outside, which is reflected by the torus hemisphericus on the inside. The protracted dorsal and bilateral evagination of the telencephalon during stage 15 results in the formation of two primordial hemispheres which are indeed completely separated from the diencephalon: the sulcus telodiencephalicus and the torus hemisphericus, which curve around the hemispheres and the foramen of Monro, respectively, have now reached the torus transversus in front of the optic stalk. Frontally, however, the telencephalon still consists of a single bulge, because a sulcus interhemisphericus is missing at that time.

The two definitive telencephalic hemispheres are established during stage 16, in which the dorsal and lateral expansions are attended with a frontal evagination of the hemispheres and, in addition, the sulcus interhemisphericus is generated. In stage 17, the expansion in the three directions mentioned is supplemented with an extension in the occipital direction. The evagination in all four dimensions thus achieved proceeds during stage 18 and the following stages. From stage 19 onwards, the caudal and frontal parts of each hemisphere expand basally, resulting in the formation of the temporal and frontal lobes, respectively. Additionally, a primordial olfactory bulb arises at the anterior end of the frontal lobe during stages 19 and 20. During stages 21, 22, and 23 the olfactory bulb gradually becomes detached from the frontal lobe, but the olfactory ventricle remains continuous with the lateral ventricle through a narrow channel. At the end of the embryonic period, defined as the end of stage 23 (Gribnau and Geijsberts 1981), the telencephalic hemispheres almost completely cover the outer diencephalon on both sides.

4.4.2 The Ventricular Ridges

Up to developmental stage 16 the entire walls of the telencephalic hemispheres are thin, whereas during that stage the anlage of the medial ventricular ridge appears in the frontobasal part of the torus hemisphericus. The ridge arises predominantly in the basal telencephalic wall but extends into the diencephalic wall in front of the optic stalk. During stage 17 the medial ventricular ridge expands considerably and thereby obscures the corresponding part of the telo-diencephalic boundary. In stage 18 these events proceed and, in addition, a second lateral ventricular ridge developes. The two ridges are completely separated by the sulcus subpallii intermedius, whereas the sulcus subpallii dorsalis marks the boundary between the lateral ventricular ridge and the thin-walled pallial anlage of the hemisphere. In stage 19 the caudal parts of the medial and lateral ventricular ridges have united into one ventricular eminence, whereas their rostral parts are still separated by the sulcus subpallii intermedius. During stages 20 through 23 the merging of the two ventricular ridges gradually proceeds frontally. Eventually, at the end of stage 23, this process is almost completed: only a vague indentation in the most frontal part of the ventricular eminence remains as a relict of the sulcus subpallii intermedius.

The progressive expansion of both ridges causes them to protrude more and more into the lumen of the lateral ventricle. The expansion of the medial ventricular ridge also leads to a narrowing of the foramen of Monro, which at the end of the embryonic period is reduced to a small cleft-like opening. As a consequence, the diencephalic part of the medial ventricular ridge, which is caudally bordered by the sulcus intraencephalicus anterior, gradually becomes isolated from the telencephalic part and develops into the preoptic region. The ventricular eminence, on the other hand, will give rise to the striatum as well as to the amygdaloid complex.

Summarizing the events during the morphogenesis of the basal forebrain of the rhesus monkey, the following conclusions can be made:

1. The medial ventricular ridge, which originates first, has a dual origin, namely partly telencephalic and partly diencephalic.
2. The lateral ventricular ridge, which arises later, has a completely telencephalic origin.
3. The two ventricular ridges, which originally develop as entirely separate entities, gradually merge into one ventricular eminence, a process which is initiated caudally and proceeds frontally.
4. The preoptic region is a derivative of the diencephalic part of the medial ventricular ridge.

A comparison of these results with the data available in the literature reveals that the morphogenesis of the basal forebrain in the rhesus monkey concurs fairly well with that in the Chinese hamster, as described by Lammers et al. (1980). Our observations are also in agreement with those of other authors (Kodama 1926; Grünthal 1952; Hewitt 1958, 1961; Brown 1967; Kahle 1969) who stated that the first-appearing elevation at the level of the foramen of Monro constitutes the primordial medial ventricular ridge, but are in contradiction to the opinion of Humphrey (1968), who considered that elevation to be the lateral ridge. Quite another view was held by some other authors (Tiedeman 1816; von Mihalkovicz 1877; Hochstetter 1919; Källén 1951; Hamilton et al. 1972) who suggested that the two ventricular ridges arise simultaneously by the division of one single elevation.

In the rhesus monkey material studied no evidence was found for the existence of a third ventricular ridge, as is consistent with the conclusion of Lammers et al. (1980) in the Chinese hamster. Other authors identified a third ridge either in early stages (His 1904; Ziehen 1906) or in later stages (Hochstetter 1919; Kodama 1926; Källén 1951).

4.5 Comparative Chronology of the Morphogenesis of the Brain

In the literature, most of the studies on the morphogenesis of the brain in primates, rodents, birds, and lower vertebrates are focused either on a limited early phase of the embryonic period (Bartelmez and Evans 1926; Bergquist 1952a, b; Bergquist and Källén 1953a, b, 1954, 1955; Vaage 1969; O'Rahilly and Gardner 1979; Müller and O'Rahilly 1980; Davignon et al. 1980) or on some subunit of the brain, for instance the diencephalon (Coggeshall 1964; Keyser 1972), the forebrain (Lammers et al. 1980), and the olfactory system (Bossy 1980). Only in man are there extensive studies published dealing with the morphogenesis of the entire brain during the whole embryonic period (von Mihalkovicz 1877; His 1904; Bartelmez and Dekaban 1962; Blechschmidt 1973;

Table 3. Developmental characteristics of the brain during stages 13–23 in man and rhesus monkey. The ages are given in days postconception and derive from O'Rahilly (1979) and Gribnau and Geijsberts (1981), respectively

Stage	Characteristics
Stage 13 man: 28–32 days rhesus monkey: 28–30 days	Posterior neuropore closed; obtuse cervical flexure; deep cranial flexure; three primary brain vesicles; seven rhombomeres; two diencephalic neuromeres (synencephalon and parencephalon posterius)
Stage 14 man: 31–35 days rhesus monkey: 30–32 days	Slight pontine flexure; rhombencephalic sulcus limitans; two rhombencephalic longitudinal zones (basal plate and alar plate); two mesencephalic neuromeres; partial telo-diencephalic boundary
Stage 15 man: 35–38 days rhesus monkey: 30–33 days	Cervical flexure > 90°; pontine flexure deepening; sulcus rhombencephalicus basalis; three rhombencephalic longitudinal zones (basal plate subdivided); cerebellar plate developing; two isthmic neuromeres; sulcus limitans mesencephalicus; sulcus mesencephalicus basalis; three mesencephalic longitudinal zones; hypothalamic cell cord; sulcus diencephalicus basalis and ventralis; three diencephalic neuromeres (plus parencephalon anterius); one longitudinal zone in the frontal diencephalon (basal); telo-diencephalic boundary complete
Stage 16 man: 37–42 days rhesus monkey: 32–34 days	Cervical flexure ca. 90°; rhombomeres vanishing; sulcus rhombencephalicus lateralis; four rhombencephalic longitudinal zones (alar plate also subdivided); decussating fibers of trochlear nerve; primordial neurohypophysial evagination; incipient medial ventricular ridge; two separate telencephalic hemispheres
Stage 17 man: 42–44 days rhesus monkey: 34–36 days	Cervical flexure < 90°; rhombomeres absent; mesencephalic neuromeres disappearing; first fibers of posterior commissure; parencephalon anterius vanished; two longitudinal zones in the frontal diencephalon (basal and dorsal); primordial epiphysial evagination; fingerlike hypophysial evagination; medial ventricular ridge; incipient lateral ventricular ridge
Stage 18 man: 44–48 days rhesus monkey: 35–38 days	Pontine flexure > 90°; mesencephalic neuromeres absent; numerous fibers of posterior commissure; distinct epiphysis; hypothalamus and subthalamus originating; one longitudinal zone (basal) in the caudal diencephalon; two separate (medial and lateral) ventricular ridges
Stage 19 man: 48–51 days rhesus monkey: 36–42 days	Pontine flexure ca. 90°; primordial plexus choroideus IV; two diencephalic neuromeres present; closure optic stalk; ventral thalamus developing; medial and lateral ventricular ridges caudally merging; olfactory fibers reach brain; primordial plexus choroideus I
Stage 20 man: 51–53 days rhesus monkey: 38–42 days	Pontine flexure < 90°; plexus choroideus IV; sulcus mesencephalicus lateralis; four mesencephalic longitudinal zones; diencephalic neuromeres absent; fasciculus retroflexus; dorsal thalamus developing; two longitudinal zones in entire diencephalon; sulcus diencephalicus dorsalis; first fibers in optic chiasm; frontal and temporal lobes of cerebral hemispheres; merging of medial and lateral ventricular ridge into one ventricular eminence proceeds frontally; primordial olfactory bulb; incipient primary cortex
Stage 21 man: 53–54 days rhesus monkey: 40–44 days	Rhombencephalic and cerebellar cell clusters; dorsal thalamus expanding caudally; hypothalamic and ventral thalamic cell clusters; numerous optic chiasm fibers; primordial plexus choroideus III; ventricular eminence expanding frontally; globus pallidus; olfactory bulb and ventricle

Table 3 (continued)

Stage	Characteristics
Stage 22 man: 54–56 days rhesus monkey: 44–48 days	Dorsal thalamus; epithalamus developing; plexus choroideus III; hypophysial lumen detached from third ventricle; putamen; amygdaloid complex; medial and lateral ventricular ridges frontally separated
Stage 23 man: 56–60 days rhesus monkey: 46–50 days	Sulcus mesencephalicus basalis absent; three mesencephalic longitudinal zones; nucleus ruber; substantia nigra; nucleus interpeduncularis; adhesio interthalamica; fasciculus mamillotegmentalis; fornix; merging of medial and lateral ventricular ridge into ventricular eminence almost completed; olfactory bulb detached from frontal lobe; first fibers of anterior commissure

Yokoh 1968; Lemire et al. 1975; O'Rahilly and Gardner 1971, 1977; O'Rahilly et al. 1982).

In addition, there are a number of investigations dealing with the staging of prenatal development in individual species in which the developmental state of the brain was directly used as a criterion in classification of the embryos, as was executed in man by Streeter (1942, 1945, 1948, 1951), O'Rahilly (1979), and O'Rahilly et al. (1981, 1982), in the rhesus monkey by Gribnau and Geijsberts (1981), and in the Chinese hamster by ten Donkelaar et al. (1979). In other staging studies the brain was casually mentioned and illustrated, e.g., by Hendrickx (1971) in the baboon, Hendrickx and Sawyer (1975) in the rhesus monkey, Phillips (1976) in the marmoset, Christie (1964) in the rat, Theiler (1972) in the mouse, and Butler (1983) in the lesser galago. A comparison of the data presented in all the studies mentioned with the results described in the present investigation leads to the conclusion that the sequence of events occurring during the morphogenesis of the brain is highly similar in rodents and primates and even identical in monkeys and man. Thus the closer the phylogenetic relationship between two species, the higher the similarity of the morphogenesis of their brains.

Another question, of course, is whether in different species the transformations constituting the morphogenesis of the brain take place at corresponding times during their development. This question, however, can only be answered for those species in which the morphogenetic data are based upon embryonic material that is classified according to accurately defined developmental stages. In a previous study the present authors (Gribnau and Geijsberts 1981), using inner and outer morphologic criteria, introduced a common staging system applicable to both rodent and primate species, as was originally developed in man by Streeter (1942, 1945, 1948, 1951) and revised by O'Rahilly (1972, 1973, 1978, 1979). On the basis of that staging system they demonstrated that a common developmental plan exists in rodents and in primates: the sequence of arisal of the individual organs is highly similar and, secondly, a mutual correlation exists between the developmental states of various organs.

Unfortunately, the morphogenesis of the entire brain in rodents has not yet been analyzed in staged embryos. From the incomplete data described in the literature mentioned above, and taking into account the modifications made earlier by the present authors (Gribnau and Geijsberts 1981) in the Chinese hamster staging system introduced by ten Donkelaar et al. (1979) and followed by Lammers et al. (1980), the tentative conclusion can be expressed that the timing of morphogenesis of the brain in rodents and primates is probably quite similar.

On the other hand, the timing of morphogenesis of the brain in human embryos is rather well known from the detailed descriptions of the authors named earlier in this section. A comparison of both their results with ours, as well as the respective developmental stages in man and rhesus monkey, leads to the conclusion that the timing of morphogenesis of the brain during the embryonic period in man and rhesus monkey is identical. A summary of the most characteristic features of the brains of both species during developmental stages 11 through 23 is given in Table 3, together with the respective ages indicated by O'Rahilly (1979) and Gribnau and Geijsberts (1981). In conclusion, it can be stated that in primates, not only the sequence but also the timing of the morphogenetic events in the developing brain is identical.

5 Summary

The morphogenesis of the brain in the rhesus monkey was studied using embryos representing organogenetic developmental stages 13 through 23 as defined earlier (Gribnau and Geijsberts 1981). At least two embryos of each stage were serially sectioned, in either the transverse or the horizontal direction. Three-dimensional reconstructions were made of the brains of embryos belonging to stages 15, 17, 19, 20, and 23. In that way a better understanding was reached of the continuously changing spatial relationships within the brain caused by the transformations occurring during its development.

In the early developmental stages 13 through 16, the morphology of the rhesus monkey brain is dominated by the transversely oriented neuromeres. A total of 7 rhombomeric, 2 isthmic, 2 mesencephalic, 3 diencephalic, and 1 telencephalic neuromeres were identified.

During further development, the neuromeric pattern is gradually replaced by a longitudinal zonal pattern: in the mesencephalon and rhombencephalon this pattern consists of four zones separated by three ventricular sulci, whereas the diencephalon exhibits two longitudinal zones bordered by three sulci. The process of changing patterns starts caudally and proceeds frontally: the rhombomeres disappearing during stage 16, the mesencephalic neuromeres during stage 17, and the diencephalic ones during stages 18 and 19. In addition to this caudal-to-rostral developmental gradient, a basal-to-dorsal gradient also exists: in the entire brain basal parts develop earlier than dorsal parts.

The rhombencephalon and the mesencephalon retain their four-zone pattern throughout development, the two basal zones being separated from the two alar zones by the sulcus limitans of His. In the diencephalon, however, the two longitudinal zones, a basal and a dorsal one, eventually become distorted because of differential growth of their subunits. Each of the three original diencephalic neuromeres is divided into a basal and a dorsal part by the sulcus diencephalicus ventralis, which ends frontally in the preoptic recess. The most caudal diencephalic neuromere, the syncncephalon, gives rise to the prerubral tegmentum basally and the pretectal area dorsally, whereas the middle diencephalic neuromere, the parencephalon posterius, develops into the subthalamus basally and the dorsal thalamus and epithalamus, dorsally. The basal part of the parencephalon anterius gives rise to the mamillary region, the hypothalamus, and the postoptic and supraoptic regions, in caudal-to-rostral order. Within the dorsal parencephalon anterius both the ventral thalamus and the preoptic region arise. The disproportionate outgrowth of the dorsal thalamus results in a lateral displacement of the ventral thalamus, which complicates the morphological pattern of the diencephalon. The original interneuromeric boundaries, however, remain visible up to stage 23 because fiber systems develop within these confines, for instance the posterior commissure and the fasciculus retroflexus, which develop within the syn-mesencephalic and the interparencephalic boundary, respectively. For this reason, the neuromeres rather than the longitudinal zones, are considered to be the primary fundamental subunits of the brain. The longitudinal zonal pattern is secondarily imposed upon the transversely oriented neuromeres due to a heterochronous histological activity of their respective basal and dorsal parts.

In the prosencephalon, the demarcation of the telencephalon from the diencephalon is completed during stage 15; two individual telencephalic hemispheres arise as bilateral evaginations of one single telencephalic neuromere during stages 13 through 16. The dorsal, lateral, and caudal expansion of the hemispheres in stages 16, 17, and 18 is followed by a frontally and basally directed outgrowth in stage 19, resulting in the formation of the frontal and temporal lobes. During stage 16 and subsequent stages in both sides of the basal forebrain a medial ventricular ridge develops, which crosses the telodiencephalic boundary: it has both a telencephalic and a diencephalic (parencephalon anterius) origin. In stage 18 a second, completely telencephalic lateral ventricular ridge arises, the two ridges being separated by the sulcus subpallii intermedius. From stage 19 onwards, the two ridges start to fuse caudally, merging into one ventricular eminence. During stages 20 through 23 this phenomenon proceeds frontally, attended with a caudal to frontal disappearance of the sulcus subpallii intermedius. At the same time the original diencephalic part of the medial ventricular ridge becomes isolated from the rest and develops into the preoptic region. At the end of stage 23 these processes are almost complete and eventually the ventricular eminence will give rise to both the striatum and the amygdaloid complex.

A comparison of the morphogenetic events during the development of the brain in the rhesus monkey, as described in the present investigation, with those in rodents and in primates, as known from the literature, leads to the conclusion that both the sequence and timing of those events are (1) similar in rodents and primates and (2) identical in rhesus monkey and man.

Acknowledgements

The authors wish to thank Prof. Dr. R. Nieuwenhuys for his valuable suggestions during the execution of this investigation and his useful comments on the manuscript. Mr. J. de Bekker is gratefully acknowledged for the preparation of the half-tone illustrations as well as the drawings. We are also indebted to Mrs. Y. van den Bos-Willemse and Mr. H. van Aanholt for their expert histotechnical assistance, to Mr. C. de Bruin for the photographs, and to Miss. A. Siebring for typing the manuscript. The maintenance, breeding, and surgery of the monkeys was skilfully conducted by Mr. A. Peters and Mr. G. Grutters at the Central Animal Laboratory of the Medical Faculty of the University of Nijmegen.

6 References

Bartelmez GW, Dekaban AS (1962) The early development of the human brain. Contrib Embryol Carneg Instn 37:13–32

Bartelmez GW, Evans HM (1926) The development of the human embryo during the period of somite formation including embryos with 2 to 16 pairs of somites. Contrib Embryol Carneg Instn 17:1–67

Bergquist H (1932) Zur Morphologie des Zwischenhirns bei niederen Wirbeltieren. Acta Zool (Stockh) 13:57–303

Bergquist H (1952a) The formation of neuromeres in homo. Acta Soc Med Ups 57:23–32

Bergquist H (1952b) Studies on the cerebral tube in vertebrates. The neuromeres. Acta Zool (Stockh) 33:117–187

Bergquist H, Källén B (1953a) Studies on the topography of the migration areas in the vertebrate brain. Acta Anat (Basel) 17:324–369

Bergquist H, Källén B (1953b) On the development of neuromeres to migration areas in the vertebrate cerebral tube. Acta Anat (Basel) 18:65–73

Bergquist H, Källén B (1954) Notes on the early histogenesis and morphogenesis of the central nervous system in vertebrates. J Comp Neurol 100:627–659

Bergquist H, Källén B (1955) The archencephalic neuromery in *Ambystoma punctatum;* an experimental study. Acta Anat (Basel) 24:208–214

Blechschmidt E (1973) Die pränatalen Organsysteme des Menschen. Hippokrates, Stuttgart

Bodian D (1936) A new method for staining nerve fibers and nerve endings in mounted paraffin sections. Anat Rec 65:89–97

Bossy J (1980) Development of olfactory and related structures in staged human embryos. Anat Embryol 161:225–236

Brown JW (1967) The development of the amygdaloid complex in insectivorous bat embryos. Ala. J Med Sci 4:399–415

Butler H (1972) The chronology of embryogenesis in the lesser galago: a preliminary account. Folia Primatol (Basel) 18:368–378

Butler H (1983) The embryology of the lesser galago (*Galago senegalensis*). Contrib Primatol 19:1–156

Christie GA (1964) Developmental stages in somite and post-somite rat embryos based on external appearance and including some features of the macroscopic development of the oral cavity. J Morphol 114:263–286

Coggeshall RE (1964) A study of diencephalic development in the albino rat. J Comp Neurol 122:241–269

Davignon RW, Parker RM, Hendrickx AG (1980) Staging of the early embryonic brain in the baboon (*Papio cynocephalus*) and rhesus monkey (*Macaca mulatta*). Anat Embryol 159:317–334

Gaunt WA (1971) Microreconstruction. Pitman, London

Gribnau AAM (1975) Immunologic pregnancy test in the rhesus monkey (*Macaca mulatta*). J Primatol 4:65–69

Gribnau AAM, Geijsberts LGM (1981) Developmental stages in the rhesus monkey (*Macaca mulatta*). Adv Anat Embryol Cell Biol 68:1–84

Gribnau AAM, Geijsberts LGM (1984) Early histogenesis of the forebrain in staged rhesus monkey embryos (to be published)

Gribnau AAM, Lammers GJ (1976) The preparation of graphical and threedimensional reconstructions of the developing central nervous system. Acta Morphol Neerl Scand 14:1–18

Grünthal E (1952) Untersuchungen zur Ontogenese und über den Bauplan des Gehirns. In: Feremutsch K, Grünthal E (eds) Beiträge zur Entwicklungsgeschichte und normalen Anatomie des Gehirns. Karger, Basel, pp 5–32

Hamilton WJ, Boyd JD, Mossman HW (1972) Human embryology. Prenatal development of form and function. Heffers, Cambridge

Hendrickx AG (1971) Embryology of the baboon. University of Chicago Press, Chicago

Hendrickx AG, Sawyer RH (1975) Embryology of the rhesus monkey. In: Bourne GH (ed) The rhesus monkey, vol II. Management, reproduction and pathology. Academic, New York, pp 141–169

Herrick CJ (1899) The cranial and first spinal nerves of Menidia: a contribution upon the nerve components of the bony fishes. J Comp Neurol 9:153–455

Herrick CJ (1910) The morphology of the forebrain in amphibia and reptilia. J Comp Neurol 20:413–547

Hewitt W (1958) The development of the human caudate and amygdaloid nuclei. J Anat 92:377–382

Hewitt W (1961) The development of the human internal capsule and lentiform nuclei. J Anat 95:191–199

His W (1888) Zur Geschichte des Gehirns, sowie der zentralen und peripherischen Nervenbahnen beim menschlichen Embryo. Abh math-phys Kl Kgl sächs Ges Wiss 14:341–372

His W (1893) Vorschläge zur Einteilung des Gehirns. Arch Anat Physiol Anat Abt 172–180

His W (1904) Die Entwicklung des menschlichen Gehirns. Hirzel, Leipzig

Hochstetter F (1919) Beiträge zur Entwicklungsgeschichte des menschlichen Gehirns. Teil I. Deuticke, Wien

Humphrey T (1968) The development of the human amygdala during early embryonic life. J Comp Neurol 132:135–165

Huxley TH (1871) The anatomy of the vertebrated animals. London

Johnston JB (1902) An attempt to define the primitive functional divisions of the central nervous system. J Comp Neurol 12:87–106

Kahle W (1969) Die Entwicklung der menschlichen Großhirnhemisphäre. Springer, Berlin Heidelberg New York

Källén B (1951) Embryological studies on the nuclei and their homologization in the vertebrate forebrain. Kgl Fysiogr Sällsk Lund Handl NF 62, nr 5

Källén B (1952) Notes on the proliferation processes in the neuromeres in vertebrate embryos. Acta Soc Med Ups 57:111–118

Källén B (1953) On the significance of the neuromeres and similar structures in vertebrate embryos. J Embryol Exp Morphol 1:387–392

Källén B (1954) On the segmentation of the central nervous system. Kgl Fysiogr Sällsk Lund Handl NF 64, nr 18

Källén B (1955) Neuromery in living and fixed chick embryos. Kgl Fysiogr Sällsk Lund Handl NF 25, nr 9

Källén B, Lindskog B (1953) Formation and disappearance of neuromery in Mus musculus. Acta Anat (Basel) 18:273–282

Keyser AJM (1972) The development of the diencephalon of the Chinese hamster. Acta Anat (Basel) 83 [Suppl 59]:1–178

Kodama S (1926) Ueber die sogenannten Basalganglien. Schweiz Arch Neurol Psychiatry 19:152–177

Kuhlenbeck H (1929a) Die Grundbestandteile des Endhirns im Lichte der Bauplanlehre. Anat Anz 67:1–51

Kuhlenbeck H (1929b) Ueber die Grundbestandteile des Zwischenhirnbauplans der Anamnier. Morphol Jb 63:50–95

Kuhlenbeck H (1930) Bemerkungen über den Zwischenhirnbauplan bei Säugetiere, insbesondere beim Menschen. Anat Anz 70:122–142

Kuhlenbeck H (1933) Bemerkungen über die theoretischen Grundlagen der Hirnmorphologie. Anat Anz 72:305–309

Kuhlenbeck H (1936) Ueber die Grundbestandteile des Zwischenhirnbauplans der Vögel. Morphol Jb 77:61–109

Kuhlenbeck H (1954) The human diencephalon. A summary of development, structure, function and pathology. Confin Neurol [Suppl] 14:1–230

Lammers GJ, Gribnau AAM, ten Donkelaar HJ (1980) Neurogenesis in the basal forebrain in the Chinese hamster (Cricetulus griseus). II. Site of neuron origin: morphogenesis of the ventricular ridges. Anat Embryol 158:193–211

Lemire RJ, Loeser JD, Leech RW, Alvord EC (1975) Normal and abnormal development of the human nervous system. Harper and Row, New York

Meek A (1907) The segments of the vertebrate brain and head. Anat Anz 31:408–415

Müller F, O'Rahilly R (1980) The early development of the nervous system in staged insectivore and primate embryos. J Comp Neurol 193:741–752

Nieuwenhuys R (1974) Topological analysis of the brain stem: a general introduction. J Comp Neurol 156:255–276

O'Rahilly R (1972) Guide to the staging of human embryos. Anat Anz 130:556–559

O'Rahilly R (1973) Developmental stages in human embryos, including a survey of the Carnegie collection. Part A: embryos of the first three weeks (stages 1 to 9). Carnegie Institution of Washington Publications, Baltimore, nr 631

O'Rahilly R (1978) The timing and sequence of events in the development of the human digestive system and associate structures during the embryonic period proper. Anat Embryol 153:123–136

O'Rahilly R (1979) Early human development and the chief sources of information on staged human embryos. Eur J Obstet Gynecol Reprod Biol 9/4:273–280

O'Rahilly R, Gardner E (1971) The timing and sequence of events in the development of the human nervous system during the embryonic period proper. Z Anat Entwickl-Gesch 134:1–12

O'Rahilly R, Gardner E (1977) The developmental anatomy and histology of the human central nervous system. In Vinken PJ, Bruyn GW, Myrianthopoulos NC (eds) Handbook of clinical neurology, vol 30. Elsevier, Amsterdam, pp 15–40

O'Rahilly R, Gardner E (1979) The initial development of the human brain. Acta Anat 104:123–133

O'Rahilly R, Bossy J, Müller F (1981) An introduction to the staging of the human embryo. Bull An Anat 65:1–98

O'Rahilly R, Müller F, Bossy J (1982) Atlas des stades du développement du système nerveux chez l'embryon humain intact. Arch Anat Histol Embryol 65:57–76

Orr HA (1887) Contributions to the embryology of the lizard. J Morphol 1:311–372

Phillips JR (1976) The embryology of the common marmoset (Callithrix jacchus). Adv Anat Embryol Cell Biol 52, nr. 5

Romeis B (1968) Mikroskopische Technik. Oldenbourg, München

Streeter GL (1942) Developmental horizons in human embryos. Description of age group XI, 13–20 somites, and age group XII, 21–29 somites. Contrib Embryol Carneg Instn 30:211–245

Streeter GL (1945) Developmental horizons in human embryos. Description of age group XIII, embryos about 4 or 5 millimeters long, and age group XIV, period of indentation of the lens vesicle. Contrib Embryol Carneg Instn 31:27–63

Streeter GL (1948) Developmental horizons in human embryos. Description of age groups XV, XVI, XVII and XVIII, being the third issue of a survey of the Carnegie collection. Contrib Embryol Carneg Instn 32:133–203

Streeter GL (1951) Developmental horizons in human embryos. Description of age groups XIX, XX, XXI, XXII and XXIII, being the fifth issue of a survey of the Carnegie collection. Contrib Embryol Carneg Instn 34:165–196

ten Donkelaar HJ, Geijsberts LGM, Dederen PJW (1979) Stages in the prenatal development of the Chinese hamster (Cricetulus griseus). Anat Embryol 156:1–28

The Boulder Committee (1969) Embryonic vertebrate central nervous system: revised terminology. Anat Rec 166:257–262

Theiler K (1972) The house mouse. Development and normal stages from fertilization to 4 weeks of age. Springer, Berlin Heidelberg New York

Tiedemann F (1816) Anatomie und Bildungsgeschichte des Gerhins im Foetus des Menschen, nebst einer vergleichenden Darstellung des Hirnbaues in den Thieren. Stein, Nürnberg

Vaage S (1969) The segmentation of the primitive neural tube in chick embryos (Gallus domesticus). Ergebn Anat Entwickl-Gesch 41:1–88

von Kupffer C (1906) Die Morphogenie des Zentralnervensystems. In: Hertwig O (ed) Handbuch der vergleichenden und experimentellen Entwicklungslehre der Wirbeltiere. Fischer, Jena

von Mihalkovicz V (1877) Entwicklungsgeschichte des Gerhirns. Nach Untersuchungen an höheren Wirbeltieren und dem Menschen. Engelman, Leipzig

Yokoh Y (1968) The early development of the nervous system in man. Acta Anat 71:492–518

Ziehen T (1906) Morphogenie des Zentralnervensystems der Säugetiere. In: Hertwig O (ed) Handbuch der vergleichenden und experimentellen Entwicklungslehre der Wirbeltiere. Fischer, Jena

Subject Index

Embryonic age 4, 5, 56
— disc 7
— length 5
— period 5, 37, 54, 55, 58
Eminence, ventricular 3, 22, 27, 32, 34, 36, 37, 38, 53, 54, 55, 56, 57, 60
Eminentia thalami 42
Epiphysis 16, 18, 26, 28, 37, 42, 56
Epithalamic region 26, 28, 32, 36, 42
Epithalamus 2, 36, 48, 49, 53, 57, 59
Evagination 1, 3, 7, 9, 11, 14, 15, 16, 18, 25, 27, 49, 53, 54, 56, 60

Fasciculus mamillotegmentalis 41, 42, 43, 57
— retroflexus 26, 28, 36, 42, 43, 46, 50, 53, 56, 59
Female 4
Fertilized egg 7
Fetus 7, 37
Fiber system 37, 41, 44, 50, 59
Film 4
Fixation 4
Flexure, cephalic 2
—, cerebral 2
—, cervical 2, 7, 9, 14, 16, 26, 27, 32, 56
—, cranial 2, 7, 9, 14, 56
—, pontine 2, 9, 11, 14, 16, 20, 22, 26, 27, 32, 34, 43, 56
Follicle(s) 42
Foramen, interventricular (Monro) 1, 11, 14, 18, 22, 24, 27, 36, 37, 38, 42, 52, 54, 55
Forebrain 1, 6, 11, 45, 53, 55, 60
Fornix 37, 41, 57
Fourth ventricle 9, 14, 16, 20, 22, 26, 32, 34, 43
Frontal lobe 26, 27, 32, 34, 37, 54, 56, 57, 60

Galago 57
Ganglion (-a) 9, 11, 20
Germ layer 1
Gestation period 4
Globus pallidus 32, 36, 37, 56
Gonadotropin, chorionic 4
Graphical reconstruction 5, 6, 42, 53

Habenular nuclei 42
Hamster, Chinese 3, 22, 45, 46, 49, 50, 55, 57, 58
Hemisphere(s), telencephalic 1, 3, 7, 9, 11, 14, 16, 18, 20, 22, 26, 32, 34, 36, 37, 53, 54, 56, 60
Heterochrony 2, 47, 60
Histogenesis 1, 34, 36, 42, 43, 44, 47, 49
Hypophysis, adeno- 15, 18, 34, 42
—, neuro- 15, 18, 25, 27, 34, 36, 37, 38, 41, 42, 56

Hypothalamic cell cord 9, 11, 13, 18, 20, 24, 25, 48, 56
— region 27, 36, 38, 41, 48, 53
Hypothalamus 2, 27, 32, 34, 38, 41, 48, 50, 56, 59
Hysterotomy 4

Infundibular recess 13, 36
Intermediate zone 37, 42, 43
Internal capsule 27, 32, 36, 37
Interneuromeric boundary 36, 50, 52, 59
Interparencephalic boundary 46, 50, 52, 59
— crest 13, 15, 18, 22, 26, 27, 53
Interventricular foramen (Monro) 1, 11, 14, 18, 22, 24, 27, 36, 37, 38, 42, 52, 54, 55
Isthmic neuromere(s) 14, 45, 56, 59
—, region 9, 14, 20, 45
Isthmus rhombencephali 9, 14, 16, 20, 26, 32, 43

Lamina terminalis 11, 18, 38
Lateral recess 32, 34, 43
— ventricle 1, 18, 22, 27, 32, 34, 36, 37, 54
— ventricular ridge 3, 18, 20, 22, 27, 32, 36, 37, 54, 55, 56, 57, 60
Length, embryonic 5
Lens 18
— lumen 18
— placode 9
— vesicle 11
Lobe, frontal 26, 27, 32, 34, 37, 54, 56, 57, 60
—, temporal 26, 32, 34, 37, 54, 56, 60
Longitudinal zone 2, 3, 9, 14, 18, 20, 26, 30, 32, 34, 43, 44, 45, 47, 48, 49, 50, 56, 57, 59, 60
Lower vertebrates 49, 55
Lumen, lens 18

Macaca mulatta 4
Male 4
Mammals 2, 7
Mamillary region 25, 27, 41, 53, 59
Man 2, 5, 56, 57, 58, 60
Marmoset 57
Mating 4, 5
Medial ventricular ridge 3, 14, 16, 18, 20, 22, 24, 27, 32, 36, 37, 38, 48, 54, 55, 56, 57, 60
Median section 5
Medulla oblongata 34
Menstrual cycle 4
Mesencephalic neuromere(s) 9, 14, 16, 20, 22, 45, 56, 59
Mesencephalon 1, 2, 7, 9, 14, 16, 20, 22, 26, 28, 30, 32, 34, 36, 42, 43, 45, 47, 56, 59

Mes-rhombencephalic boundary 9, 20
Metencephalon 1, 26, 32, 43
Methylbenzoate 4, 7, 9, 16, 20, 22, 27, 32, 34, 37
Microscopic section 4, 5, 7, 9, 11, 14, 16, 18, 20, 27, 38, 42, 59
Migration area 2
–, radial 47
Mitotic activity 2
Morphogenesis 1, 2, 3, 7, 34, 36, 45, 47, 53, 55, 57, 58, 59, 60
Morphology 2, 9, 13, 14, 18, 20, 22, 26, 27, 34, 36, 38, 42, 43, 44, 48, 49, 59
Motor column 2
Mouse 57
Myelencephalon 1, 26, 27, 32, 34, 43, 47

Nasal epithelium 22
Nerve, cranial 9, 11, 27, 32, 43, 44
–, trochlear 16, 20, 26, 32, 56
Neural groove 1
Neural plate 1, 7
– tube 1, 2, 7, 11, 45, 53
Neuraxis 2, 7
Neuroepithelium 11
Neurohypophysis 15, 18, 25, 27, 34, 36, 37, 38, 41, 42, 56
Neuromere(s) 2, 3, 9, 45, 49, 53, 59, 60
–, diencephalic 9, 13, 14, 16, 18, 20, 22, 26, 27, 28, 45, 46, 47, 49, 50, 53, 56, 59
–, isthmic 14, 45, 56, 59
–, mesencephalic 9, 14, 16, 20, 22, 45, 56, 59
–, optic 46, 49, 50
–, parencephalic 13, 14, 46
–, postoptic 46, 50
–, prosencephalic 45, 49, 50
–, rhombencephalic 9, 10, 11, 14, 16, 20, 45, 56, 59
–, secondary 2
–, synencephalic 13, 50
–, telencephalic 45, 49, 50, 59, 60
–, tertiary 2
Neuromeric pattern 3, 46, 47, 49, 50, 59
Neuromerism 45, 46, 47
Neuropore, anterior 1, 7
–, posterior 1, 7, 56
Nucleus (-i) 2, 36
– Darkschewitsch 43
– habenularis 42
– hypothalamicus posterior 41
– interpeduncularis 43, 57
– interstitialis 43
–, mamillary 41
– olivaris 44
– posterior thalami 43
–, pretectal 43
– ruber 43, 57
– ventromedialis hypothalami 38

Olfactory bulb 22, 27, 32, 36, 37, 54, 56, 57
– fibers 22, 27, 56
– placode 9
– region 32
– system 55
– ventricle 32, 36, 37, 54, 56
Ontogenesis 1, 2, 14, 49
Optic chiasm 25, 27, 34, 38, 56
– cup 9, 11, 18
– evagination 14
– fissure 18
– neuromere(s) 46, 49, 50
– recess 24, 27
– region 46
– stalk 9, 11, 15, 18, 24, 54, 56
Organogenetic period 5
– phase 5, 7
– stage 7, 56, 59
Orthogonal projection 5
Otic vesicle 9, 11, 45
Ovulation 6

Palate, secondary 7, 37
Pallial region 20, 27, 36, 37, 54
Parencephalic neuromere(s) 13, 14, 46
Parencephalon anterius 13, 14, 18, 22, 24, 27, 36, 41, 42, 46, 48, 50, 52, 53, 56, 59, 60
– posterius 9, 13, 14, 16, 18, 22, 26, 27, 28, 32, 36, 42, 43, 46, 48, 49, 50, 53, 56, 59
Pars distalis (hypophysis) 42
– dorsalis thalami 2
– ventralis thalami 2
Pattern, neuromeric 3, 46, 47, 49, 50, 59
–, zonal 2, 30, 43, 44, 47, 48, 49, 50, 59, 60
Pedunculus cerebri 43
Period, embryonic 5, 37, 54, 55, 58
–, gestation 4
–, organogenetic 5
Phase, organogenetic 5, 7
–, postsomite 5
–, presomite 5
–, somite 5
Photography 4
Placode, lens 9
–, olfactory 9
Plate, alar 2, 16, 20, 22, 26, 30, 32, 43, 44, 47, 56, 59
–, basal 2, 16, 20, 22, 26, 28, 30, 32, 43, 44, 47, 56
–, cerebellar 20, 22, 26, 27, 32, 34, 56
–, neural 1, 7
Plexus choroideus 22, 26, 27, 28, 32, 34, 38, 43, 56, 57
Plica encephali ventralis 2
Polystyrene model 6

67

Advances in Anatomy, Embryology and Cell Biology

Editors: F. Beck, W. Hild, R. Ortmann,
J. E. Pauly, T. H. Schiebler

Volume 88
C. Schulze

Sertoli Cells and Leydig Cells in Man

1984. 41 figures. VI, 106 pages
ISBN 3-540-13603-7

Volume 87
D. B. Weishampel

Evolution of Jaw Mechanisms in Ornithopod Dinosaurs

1984. 20 figures. VIII, 110 pages
ISBN 3-540-13114-0

Volume 86
J. A. Winer

The Medial Geniculate Body of the Cat

1984. 45 figures. Approx. 110 pages
ISBN 3-540-13254-6

Volume 85
J. Altman, S. A. Bayer

The Development of the Rat Spinal Cord

1984. 126 figures. VIII, 166 pages
ISBN 3-540-13119-1

Volume 84
F. Hajós, E. Bascó

The Surface-Contact Glia

1984. 25 figures. VI, 81 pages
ISBN 3-540-13243-0

Volume 83
W. K. Schwerdtfeger

Structure and Fiber Connections of the Hippocampus

A Comparative Study
1984. 40 figures. VI, 74 pages
ISBN 3-540-13092-6

Volume 82
H. Scheich, S. O. E. Ebbesson

Multimodal Torus in the Weakly Electric Fish Eigenmannia

1983. 39 figures. VII, 69 pages
ISBN 3-540-12517-5

Volume 81
U.-F. Habenicht, F. Neumann

Hormonal Regulation of Testicular Descent

1983. 39 figures. VI, 55 pages
ISBN 3-540-12439-X

Volume 80
J. Koebke

A Biomechanical and Morphological Analysis of Human Hand Joints

1983. 50 figures. VI, 85 pages
ISBN 3-540-12438-1

Volume 79
S. F. Perry

Reptilian Lungs

Functional Anatomy and Evolution
1983. 32 figures. VII, 81 pages
ISBN 3-540-12194-3

Springer-Verlag
Berlin
Heidelberg
New York
Tokyo

R.V.Krstić

Illustrated Encyclopedia of Human Histology

1984. 1576 figures. Approx. 465 pages. ISBN 3-540-13142-6

The astounding progress in histologic investigation over the last three decades and the often confusing array of synonyms for important new structures and morpho-functional phenomena it has generated make Professor Krstić's **Illustrated Encyclopedia of Human Histology** especially timely. In it he provides practicing histologists, clinicians, students, and medical researchers with an authoritative reference work which defines and illustrates fundamental histologic entities and processes while enumerating variant terminology.

The book's coverage embraces cell biology, general and special human histology, histophysiology, and where relevant, embryology and veterinary histology. The items are arranged alphabetically, with synonyms and Latin terms according to Nomina Anatomica (1977), if such exist, indicated in parentheses. These are followed by a brief definition and one or more illustrations. All items are extensively cross-referenced and conclude with additional information as well as up-to-date lists of pertinent literature.

R.V.Krstić

Ultrastructure of the Mammalian Cell

An Atlas

With a Foreword by W. Bargmann
Translated from the German by A. R. von Hochstetter
1979. 176 plates drawn by the author. XV, 376 pages.
ISBN 3-540-09583-7

From the reviews:
"It is seldom that one reviews a book which provides so much concentrated visual pleasure and which, at the same time, has more than succeeded as an instructive and readable textbook for students engaged in the biological sciences."   *(Journal of Audiovisual Media in Medicine)*

"One could continue endlessly pointing out the very extensive range of this atlas but the only answer is to acquire it oneself whether for the library or the laboratory, it cannot fail to be of use."

(Proceedings of the Royal Microscopical Society)

Springer-Verlag
Berlin
Heidelberg
New York
Tokyo